Therese Helland

Software Process Support

Therese Helland

Software Process Support

A Service-Oriented Approach

VDM Verlag Dr. Müller

Impressum/Imprint (nur für Deutschland/ only for Germany)
Bibliografische Information der Deutschen Nationalbibliothek: Die Deutsche Nationalbibliothek
verzeichnet diese Publikation in der Deutschen Nationalbibliografie; detaillierte bibliografische
Daten sind im Internet über http://dnb.d-nb.de abrufbar.
Alle in diesem Buch genannten Marken und Produktnamen unterliegen warenzeichen-, marken-
oder patentrechtlichem Schutz bzw. sind Warenzeichen oder eingetragene Warenzeichen der
jeweiligen Inhaber. Die Wiedergabe von Marken, Produktnamen, Gebrauchsnamen,
Handelsnamen, Warenbezeichnungen u.s.w. in diesem Werk berechtigt auch ohne besondere
Kennzeichnung nicht zu der Annahme, dass solche Namen im Sinne der Warenzeichen- und
Markenschutzgesetzgebung als frei zu betrachten wären und daher von jedermann benutzt
werden dürften.

Coverbild: www.purestockx.com

Verlag: VDM Verlag Dr. Müller Aktiengesellschaft & Co. KG
Dudweiler Landstr. 99, 66123 Saarbrücken, Deutschland
Telefon +49 681 9100-698, Telefax +49 681 9100-988, Email: info@vdm-verlag.de

Herstellung in Deutschland:
Schaltungsdienst Lange o.H.G., Berlin
Books on Demand GmbH, Norderstedt
Reha GmbH, Saarbrücken
Amazon Distribution GmbH, Leipzig
ISBN: 978-3-639-09616-3

Imprint (only for USA, GB)
Bibliographic information published by the Deutsche Nationalbibliothek: The Deutsche
Nationalbibliothek lists this publication in the Deutsche Nationalbibliografie; detailed
bibliographic data are available in the Internet at http://dnb.d-nb.de.
Any brand names and product names mentioned in this book are subject to trademark, brand or
patent protection and are trademarks or registered trademarks of their respective holders. The use
of brand names, product names, common names, trade names, product descriptions etc. even
without a particular marking in this works is in no way to be construed to mean that such names
may be regarded as unrestricted in respect of trademark and brand protection legislation and
could thus be used by anyone.

Cover image: www.purestockx.com

Publisher:
VDM Verlag Dr. Müller Aktiengesellschaft & Co. KG
Dudweiler Landstr. 99, 66123 Saarbrücken, Germany
Phone +49 681 9100-698, Fax +49 681 9100-988, Email: info@vdm-publishing.com

Copyright © 2008 VDM Verlag Dr. Müller Aktiengesellschaft & Co. KG and licensors
All rights reserved. Saarbrücken 2008

Printed in the U.S.A.
Printed in the U.K. by (see last page)
ISBN: 978-3-639-09616-3

Software Process Support

A Service-Oriented Approach

Therese Helland

Abstract

Managing software development projects becomes increasingly cumbersome as software systems become more complex, teams become larger and more distributed, and required resources reside in different locations. By providing tools for process modelling and computerised enactment support, research on software process technology aims to aid the management and execution of software development processes. This research area is currently receiving increased attention, and the computerised support for software development is currently a 'hot topic' within the software engineering field. Several academic and commercial process support tools have been built to date. Common deficiencies include limited support for distributed work environments, little possibility for distribution of system components and insufficient integration with third-party tools.

This thesis presents recent research in the software process area by describing the design, implementation, and evaluation of the IMÅL prototype process management and support environment. The aim of the prototype is to address some of the weaknesses of existing systems. IMÅL consists of a simple process modelling language and tool, as well as a more sophisticated process engine employing some novel concepts in the software process technology area. An underlying service-oriented architecture is the basis for the engine, which is built as a collection of distributed Web Services with the intent of providing improvement in scalability and stability over previous approaches. Furthermore, multiple processing instances with different underlying paradigms are combined into one single engine, in order to utilise the advantages of several approaches. Other essential features of the prototype system include a to-do list facility and seamless integration with third-party tools.

Feedback from users involved in an evaluation survey shows that the system was favourably received, and indicates that further development of the prototype is feasible. Based on the lessons learnt from the development of the IMÅL prototype system, it is advised that future development further explore the usefulness of the distributed, service-oriented, multiple processing instances approach employed in this project. Furthermore, support for process evolution, cooperative activities, different user-types, and incorporation of time concepts, should also be investigated.

Acknowledgements

Firstly, I would like to express my sincere gratitude to my supervisors, Professor John Grundy and Professor John Hosking, whose insightful ideas, enthusiasm and support for this project have been invaluable.

I also want to acknowledge everyone else who contributed with knowledgeable input and assisted the process in any way. Thanks especially to those who participated in the survey of the IMÅL prototype system, providing useful feedback and helpful suggestions for future improvement. Also, thanks to Rowland Burdon for proof-reading the thesis and providing valuable recommendations and comments.

Furthermore, I would like to thank my loving family for always being there for me. Your support means everything to me.

Thanks also to all my friends, who never give up on me, no matter if I'm near or far away.

Finally, special thanks to my very best friend, for being a constant source of guidance and encouragement, and for making me smile.

Table of Contents

XI

List of Figures

List of Tables

List of Acronyms

BPR Business Process Reengineering

CASE Computer Aided Software Engineering

CD Cognitive Dimensions

CE Concurrent Engineering

CIM Computer Integrated Manufacturing

CSCW Computer Supported Cooperative Work

DBMS Database Management System

DTD Document Type Definition

EM Enterprise Modelling

EVPL Extended Visual Planning Language

HTML HyperText Markup Language

HTTP Hypertext Transfer Protocol

IEEE Institute of Electrical and Electronics Engineers

ISE Information Systems Engineering

JAXB Java Architecture for XML Binding

JSP Java Server Pages

JWSDP Java Web Services Developer Pack

LAN Local Area Network

MILOS Minimally Invasive Long-term Organizational Support

OD Organisational Design

OOA Object-Oriented Analysis

OOD Object-Oriented Design

PML Process Modelling Language

PPML Pounamu Process Modelling Language

PPMT Pounamu Process Modelling Tool

PSEE Process-centred Software Engineering Environments

RUP Rational Unified Process

SE Software Engineering

SGML Standard Generalized Markup Language

SOAP Simple Object Access Protocol

UDDI Universal Description, Discovery and Integration

UML Unified Modelling Language

URL Uniform Resource Locator

VPL Visual Process Language

WAN Wide Area Network

W3C World Wide Web Consortium

WfMC Workflow Management Coalition

WFMS Workflow Management System
WM Workflow Management
WSDL Web Services Description Language
XML eXtensible Markup Language
XP Extreme Programming
XSL eXtensible Stylesheet Language

Chapter 1 - Introduction

"Writing this sort of report is like building a big software system. When you've done one you think you know all the answers and when you start another you realize you don't even know all the questions."

Brian Randell [73].

This chapter introduces the topic, purpose and layout of this thesis. Firstly, it outlines the motivation behind the work enclosed in the project, and the context in which it was undertaken. Key objectives of the project are then identified, followed by an overall description of the approach adopted. Finally, an overview of the ensuing chapters of the thesis is presented.

1.1. Motivation

The planning and execution of work is an important aspect of any business environment. For decades, this has been managed manually by humans filling roles such as project manager or supervisor. Recently, process management systems have become popular in many environments like office automation and software development [42]. Software process technology, which is a fast-growing discipline in the area of information technology, allows the planning and execution of software development processes to be managed and supported by a computer program. The program assigns work to humans or other programs, passes it on, and tracks its progress. A major advantage of such systems is the increased efficiency of development processes resulting from the automation of what was previously manual and time-consuming tasks.

Various academic prototypes, as well as some commercially available products, have been developed in the category of process management and support. Examples are Regatta [88], SPADE-1 [5], ProcessWeaver [31] and EPOS [49]. Most of the existing systems are, however, limited by several deficiencies. For example, many systems deploy centralised client-server architectures in which a central server caters for process modelling and a centralised process engine provides computerised process enactment support [39]. Disadvantages of such an approach include reduced robustness and performance, and inadequate support for distributed work environments.

Although adequate support for activities such as collaboration, coordination and communication is essential in order to accommodate today's dispersed work teams and companies, another frequent shortcoming of process support systems to date is the limited support for cooperative activities. Other weaknesses, such as hard-to-understand process modelling mechanisms and insufficient integration with third-party tools, are also common [43]. Finally, most existing systems employ a homogeneous process engine, meaning that the process engine constitutes one single processing instance for enactment processing. The problem with an approach of this sort is that all processing is entrusted to one processing instance, reducing the robustness and flexibility of the engine.

This thesis report describes the design and implementation of a new process management and support tool, IMÅL, which is aimed at addressing some of the shortcomings of previous approaches. IMÅL comes from the Norwegian 'i mål', which loosely translates to 'reaching a goal', and was chosen as the name for the system because the reaching of goals is a primary purpose of both processes and the technology that supports them.

1.2. Objectives

The overall objective of this thesis project is to address some of the weaknesses of existing software process support systems. Several sub-goals have been identified. A major goal, and a novel feature of the work, is the employment of a distributed, service-oriented architecture to the design and implementation of a process support tool. Attention is primarily focused on the development of a novel, heterogeneous process engine composed of a distributed collection of services, and including several processing instances. The process modelling formalism and tool should be relatively simple, but allow for modelling of the necessary process elements. Furthermore, facilities for users to access process-related information, such as process state and progress, should also be provided by the system.

Another important objective is to provide good integration with third-party tools. This should be done by presenting users with tools they need in order to perform work on processes, as well as utilising existing tools to carry out processing tasks on behalf of the system. Additionally, the system should to be built in such a way that it can conveniently accept the plugging in of additional components in the future, such as facilities to support cooperative activities. A final aim of this work is to provide an evaluation of the proposed system, with the intent of assessing the relevance of the approach employed.

Based on the above, the following list summarises the major goals of this thesis work. As mentioned, the overall objective is to address some weaknesses of existing software process support systems through the development of a new prototype tool. The prototype and the work in general therefore aim to provide:

- A distributed and service-oriented architecture.
- A heterogeneous process engine employing multiple processing instances.
- A simple process modelling language and tool.
- Facilities for users to access process related information.
- Good integration of third-party tools.
- Openings for the plugging in of additional components.
- An evaluation of the proposed system.

1.3. Approach

In order to provide a context for this thesis work, a comprehensive literature review of related fields and previous work in the software process area is initially undertaken. Consequently, detailed requirements for the IMÅL prototype process support system are identified and elaborated on. Furthermore, designing the system's architecture is a very important task, as the goal is a completely service-oriented architecture. Because of their

23

interoperability and service-oriented nature, Web Services have been chosen as the primary building blocks for the system.

The Pounamu meta-CASE tool is used to create a simple, visual process modelling language and tool, and Pounamu is also assigned handling of user interaction in IMÅL. Furthermore, a pluggable enactment engine, providing functionality to interpret the process models and allow for computerised process enactment support, is presented. The engine is distributed into several smaller Web Services with individual responsibilities, including a simple event-based processing instance. Information about the state and progress of processes, as well as task allocation, is made available to certified users via a Web-based to-do list facility.

Subsequently, an evaluation has been carried out on the initial IMÅL prototype. The evaluation involves a Cognitive Dimensions evaluation, as well as a survey engaging potential users. Observations and feedback from the evaluation are then used as the basis for a second development increment to improve the overall system. In this increment, an additional rule-based automatic processing instance is introduced, third-party tools such as Microsoft Infopath and Idiom Decision Suite are integrated, and the to-do list service is improved to allow users to enact processes directly from within the service.

1.4. Organisation of the Thesis

The following outlines the contents of each of the subsequent chapters of this thesis:

Chapter 2 – Background and Related Work: In this chapter, the background for this thesis work is presented by investigating previous and current work in the software process area and other closely related areas.

Chapter 3 – Requirements: This chapter identifies and analyses essential requirements for the IMÅL process management and support environment.

Chapter 4 – Design: Based on the requirements analysis, this chapter presents a more detailed design of the IMÅL system, including system architecture and composition, as

24

well as design patterns used.

Chapter 5 – Implementation: This chapter is concerned with the implementation of the design into a fully working process support system, and provides details about the technologies and techniques employed.

Chapter 6 – Evaluation: In this chapter, a Cognitive Dimensions evaluation of IMÅL is presented, along with an evaluation survey carried out on the functionality and usability of the prototype.

Chapter 7 – Integrating Existing External Systems: This chapter describes the integration of two existing external systems, Microsoft Infopath and Idiom Decision Suite, into the IMÅL environment with the intention of addressing some of the weaknesses identified in the evaluation.

Chapter 8 – Conclusion and Future Work: This chapter concludes the thesis by summarising the major contributions of the work undertaken and providing suggestions and possible directions for future work.

Chapter 2 - Background and Related Work

2.1. Introduction

A variety of related fields comprise the background for this thesis research. In this chapter, these fields are defined and explained, with the intention of providing a context for the work that has been undertaken. The importance of each field to the area of interest and the interrelationships between them are elaborated on.

2.2. Software Engineering

Software engineering is a key theme of this thesis work. This is because it deals with the development of software, or in other words, the software development process. New technology is currently emerging that can be utilised in software development in order to improve the overall execution of development processes. The improvement is typically caused by more efficient communication, automation of previous manual and time-consuming tasks, and continuous user support throughout the life of a process.

The term "software engineering" traces back to a seminal NATO conference held in 1968, in which the need for an engineering approach to the development of software was addressed [62]. According to Naur and Randell, the conference was to "shed further light on the many current problems in software engineering, and also to discuss possible techniques, methods and developments which might lead to their solution" [62]. Another purpose of the conference was to identify necessities, shortcomings and trends, and use the findings as a guide for computer manufacturers and users [62].

"The phrase 'software engineering' was deliberately chosen as being provocative, in implying the need for software manufacture to be based on the types of theoretical foundations and practical disciplines, that are traditional in the established branches of engineering."[62].

Together with its sequel conference in 1969 [73], the conference propelled the development of reliable methods for software development, as a response to the chronic failure of large software projects with respect to meeting schedules and budgetary constraints. The field has developed rapidly ever since, and is currently an important and highly influential research area.

Software engineering is defined in IEEE (Institute of Electrical and Electronics Engineers) Standard 610.12 as "the application of a systematic, disciplined, quantifiable approach to the development, operation, and maintenance of software; that is, the application of engineering to software" [48]. Thus, software engineering encompasses all the activities associated with the whole life cycle of a software system.

A common belief that better technology alone can solve the software engineering problem has changed in the last few decades, and the importance of recognising software development as a business process in itself is now widely accepted. As a consequence, considerable attention is now being paid to process issues within the software engineering field. Additionally, the definition of software engineering above emphasises discipline and control, making the area suitable for the application of software process models and technology.

2.3. Software Process

Software process is a term frequently used about the steps involved in the development and maintenance of software. Thus, a software process can be seen as a collection of the steps involved in software engineering.

The concept of software process was initially introduced as 'the programming process' by Manny Lehman in 1969 [52]. His study of the IBM programming process, undertaken at the IBM Research Centre at Yorktown Heights, led him to identify the phenomenon of program evolution. In describing the results of his study, Lehman defined the programming process as "the total collection of technologies and activities that transform the germ of an idea into a binary program tape" [52]. This is seen by many as the start of the software process research area, and Lehman and others have continued further investigation into the field [53].

The first conference addressing the software process was held in England in 1984 [71], and has since been followed by many subsequent conferences devoted to the software process and related research areas.

Today, the Workflow Management Coalition (WfMC) defines a business process as "a set of one or more linked procedures or activities which collectively realise a business objective or policy goal, normally within the context of an organisational structure defining functional roles and relationships" [109]. Building on this definition of a business process, a software process can be seen as a business process where the business objective is the development and maintenance of software. In other words, the software process defines the way in which software development is organised, managed, measured, supported and improved [61].

Considering the definitions of software process and software engineering above, key links between the two areas can be readily identified. Software processes can be used in software engineering as a means to achieve the goal of a "systematic, disciplined and quantifiable approach to the development, operation, and maintenance of software", as identified by the IEEE [48]. For this reason, the interest in and focus on the software process have increased. Curtis stated that the software production process is amongst the issues that have received most attention from the software engineering community recently [26]. The number of papers published and the number of conferences and workshops held addressing the topic support this statement. Totland and Conradi identified the reason behind this increased interest as being the desire to improve software production, in terms of increased quality and reduced time and cost [92]. In other words, improving the software engineering cycle is an important goal which can be achieved by planning, controlling, monitoring and re-using the software engineering process.

Montangero et al. argued that the software process, like a manufacturing process, is composed of two interrelated processes, namely the production process and the management process [61]. The production process deals with the production and maintenance of the product, whereas the management process controls the production process and provides the resources needed. This composite view of a software process is depicted in Figure 2-1.

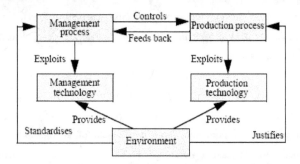

Figure 2-1. The Composite Software Process (from [29]).

2.4. Software Life Cycle and Process Models

A software life-cycle model can be seen as a description of the software development process [76]. It is constructed in order to study and understand organisations and their systems, and to aid development. Software life-cycle models can be categorised into descriptive and prescriptive models. Descriptive models describe how a completed development process has been undertaken, and are good for understanding and improving processes, while prescriptive models serve as guidelines or frameworks for how future development processes should be carried out [76].

Scacchi argued that software process models, in contrast to software life-cycle models, "often represent a networked sequence of activities, objects, transformations, and events that embody strategies for accomplishing software evolution" [76]. He further explains that process models are used in the development of more accurate and formalized descriptions of software life-cycle phases, and that "their power emerges from their utilization of a sufficiently rich notation, syntax, or semantics, often suitable for computational processing" [76].

According to Scacchi, software process models can be descriptive; characterising events that occur when people try to follow a planned chain of actions, or prescriptive; capturing an idealised plan of actions [76]. Ideally, a software process model should be of such a nature that it can be taught, explained, measured, studied and improved, in order to enhance the software process and the overall software development activity.

Despite Scacchi's distinction between software life-cycle models and software process models, the terms are commonly used to address the same concept, namely the description of a software development process. There are, however, several different views on the software development process and what aspects should be included in it. Software life-cycle and process models are constructed according to these different view-points, leading to a variety of resulting models. Some of the most common prescriptive models are introduced in the following. The Extreme Programming discipline is also briefly introduced, as it is strongly linked to the other models described.

Although differently emphasised, the following six activities are typically included in different software life-cycle and process models [14, 75]:

1. **Requirements Analysis** – identifying functional and non-functional requirements.
2. **Design** – determining overall and detailed design of the system.
3. **Implementation** – coding the system.
4. **Evaluation** – testing the system and its functionality.
5. **Deployment/Release** – deploying/releasing the system to users.
6. **Maintenance** – maintaining the system by fixing bugs and updating functionality.

2.4.1. The Waterfall Model

Winston Royce's paper of 1970, "Managing the Development of Large Software Systems" [75], is commonly regarded as the progenitor of the waterfall model. The model is characterised by a linear development process, passing through the phases of analysis, design, implementation, testing, and operation and maintenance. In other words, the model enacts the most essential activities in software development in their most primitive sequence of dependency. Each phase in the model results in a particular product, and the model has its name from the logical flow of these products from one phase to the next.

Depicted in Figure 2-2, the waterfall model usually considers the software process non-iterative. It is based on an economics-oriented approach to software development processes [13], and was primarily intended to help structure staff and manage large development projects in complex organisations [12].

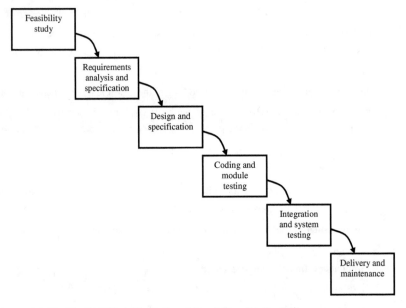

Figure 2-2. The Waterfall Model for Software Development (after [75]).

The waterfall model has often been criticised for its inflexibility, because once one phase of the model is completed and the next is initiated, it is not possible to go back and adjust any decisions made in the preceding phase. For example, once the design phase has started, the developer cannot go back and modify any decisions made in the analysis phase. Also, risk is pushed forward in time, making it costly to undo mistakes from earlier phases. Despite the criticism, however, the model is still one of the most commonly used software process models.

2.4.2. The Spiral Model

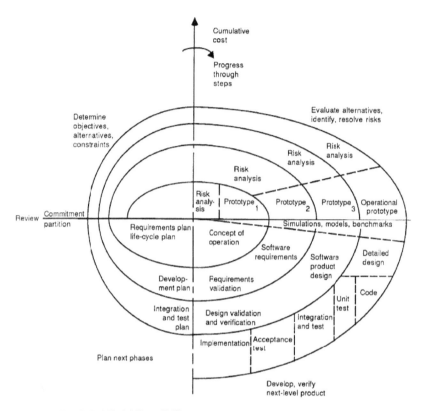

Figure 2-3. The Spiral Model (from [14]).

Illustrated in Figure 2-3, Boehm's spiral model is a refinement of the waterfall model, explicitly recognising that software development is a cyclic activity [14]. The entire waterfall model is executed once for each cycle in the spiral model, in order to achieve evolutionary development. Additionally, considerable focus is put on aiding risk management throughout the development process. When using this model, the entire system is not defined in detail at the start of the development process as when using the waterfall model. Conversely, features are defined, implemented and tested according to priority, and the feedback gathered from the testing is used as a basis for the next iteration of the model. Thus, each cycle produces a prototype which eventually results in the final system.

"The primary advantage of the spiral model is that its range of options allows it to accommodate the best features of existing software process models, while its risk-driven approach helps it to avoid most of their difficulties. In appropriate situations, the spiral model becomes equivalent to one of the existing process models. In other situations, it provides guidance on the best mix of existing approaches to be applied to a given project." [14].

The spiral model is commonly used by proponents of object-oriented design, who often perceive the waterfall model as being old-fashioned and inflexible.

2.4.3. The Evolutionary Model

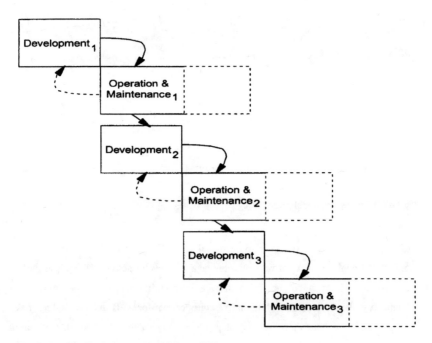

Figure 2-4. The Evolutionary Model (from [30]).

Depicted in Figure 2-4, the evolutionary model is an iterative model that focuses on prototyping. In his book about software engineering, Sommerville described evolutionary prototyping as "an approach to system development where an initial prototype is produced and refined through a number of stages to the final system" [78]. A system is developed as a series of increments, where the result of each increment is delivered to the customers. Each development increment progresses through the phases of analysis, design, implementation and testing, followed by operation and maintenance. The prototype is then deployed to customers, whose feedback on its use is incorporated into the requirements for the subsequent increment. By repeatedly addressing the requirements best understood by developers and customers at certain points in the development process, the overall aim of evolutionary prototyping is the delivery of a complete, working system [78].

Advantages of the evolutionary model include early delivery and deployment of a prototype system, as well as close user involvement in the development process [78]. On the other hand, because of its composite and broad approach, the model suffers from management and maintenance problems [78]. Generally, the evolutionary model is beneficial for the development of systems for which a complete specification cannot be defined in advance.

2.4.4. The Incremental Model

The incremental model also focuses on prototyping. However, by addressing the weaknesses of the evolutionary model, the incremental model intends to combine advantages of prototyping with a more manageable process and a better system structure [78]. Small increments and releases are planned in detail, and an overall architecture for the total system is established before the system is incrementally developed and delivered [78]. At each increment, new requirements and specifications may be developed, and the result of each increment is the inclusion of additional or improved functionality in the overall system. As an example, the Rational Unified Process (RUP) is a widely used software engineering process and framework that adopts an incremental approach to the development of software [51].

A perceived benefit of the incremental development model is that the increments are usually relatively small and well-planned, making them easier to understand, test, manage and maintain. The model, illustrated in Figure 2-5, is good for projects where user requirements are likely to change over time, as new and changing requirements can be satisfied in succeeding increments of the release plan.

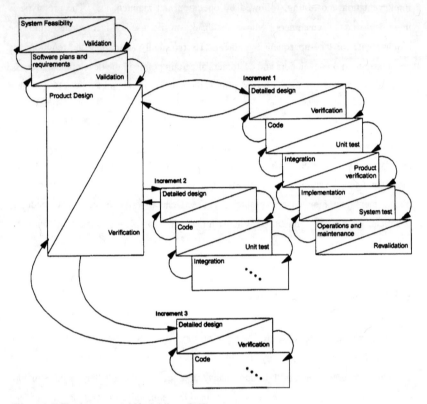

Figure 2-5. The Incremental Model (from [13]).

2.4.5. Extreme Programming

Extreme Programming (XP) is a recently developed software development methodology in which extensive emphasis is put on highly incremental and user-centred development. Initiated by Kent Beck, XP's disciplined approach to software development promises to "reduce project risk, improve responsiveness to business changes, improve productivity throughout the life of a system, and add fun to building software in teams all at the same time" [8]. XP relies on simplicity, programming in pairs and ample unit testing, and includes managers and customers as well as developers in the development process. User stories serve as the basis for development processes, and continuous feedback from customers throughout the process is intended to motivate code improvement.

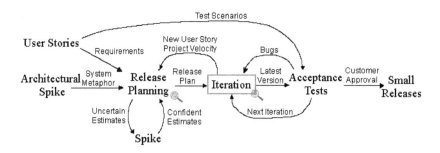

Figure 2-6. An Extreme Programming Project (from [106]).

An example of an XP project is depicted in Figure 2-6. Based on user stories and a proposed architecture, a release plan is created during a release planning meeting. Spike solutions, which are simple programs built to explore potential solutions, can be constructed to aid in design. An iterative development approach is then employed, and each iterative increment includes testing and releasing the new version of the system [106].

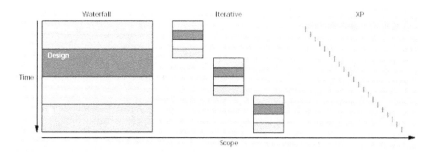

Figure 2-7. A Comparison of the Waterfall, Iterative and Extreme Programming Models (from [7]).

According to Beck, XP is designed for use by small teams developing software rapidly in an environment where requirements are vague or frequently changing [8]. Figure 2-7 shows a comparison of the classical waterfall model, iterative models such as the spiral, evolutionary and incremental models, and the XP model.

2.5. Software Process Patterns

Along with analysis and design patterns, software process patterns are among the many 'flavours' of software development patterns. In short, a software process pattern describes a proven successful approach to the development of software, and includes tasks and actions performed throughout the development process.

The concept of process patterns, and of organizational patterns in general, was first introduced by Coplien in 1994 [24]. In one of the key first papers on this topic, he presented common patterns from successful organisations, and furthermore stated that "these patterns are missing from organizations that are less productive or less successful" [24]. The paper basically serves as a guide to common successful patterns for companies to adopt in order to achieve success.

Ambler described a process pattern as "a collection of general techniques, actions, and/or tasks for developing object-oriented software" [2]. The emphasis of a process pattern should be on describing what to do, but not exactly how to do it. Ambler further argued that by applying them in an organised manner, process patterns can be used to build software processes for an organisation [1].

The following three distinct process pattern types are identified by Ambler [2, 3]:

1. **Task process patterns** – represent the steps involved in performing a single task.
2. **Stage process patterns** – capture the steps involved in completing a process stage, and can often be composed of several task process patterns.
3. **Phase process patterns** – are performed in serial order and characterize interactions between iterative stage process patterns.

From Ambler's descriptions above, it is evident that process patterns are closely related to software process models, but offer a more fine-grained partitioning of the overall process into smaller steps than do software life-cycle models. Process patterns as well as process models deal with everything down to the simple task-level, whereas process life-cycle models deal with software engineering at the phase-level.

2.6. Software Process Technology

Software process technology typically refers to software programs that aim to improve software development processes by offering efficient management of processes and computerised support for their execution. Means to both define and enact processes are therefore essential building blocks in software process technology.

In 1987, Leon Osterweil initiated the research on software process technology when he introduced a whole new concept by stating that 'software processes are software too' [68]. Osterweil argued that the concepts of software and software process are indeed related, and that the software process itself can be used as the basis for software. For that reason, he suggested that a software process development discipline be initiated based on that of traditional software development. Some formalism should be used to model software processes and tailor them to the needs of different organisations, and the resulting process models should be represented by executable code. Execution of the code would provide support and supervision of ongoing processes.

Osterweil's call for a software process development discipline based on the traditional software development discipline is of great importance for the existence of today's software process technology field. His early efforts triggered the research community's interest in this area, and a considerable amount of work has since been carried out within this field [69].

Software process technology is currently an advancing research area of considerable importance within the software engineering field. Flavio Oquendo describes how the technology aims at supporting the software engineering process by providing means to model, analyse, improve and enact software processes [66]. He further explains that recently, software process technology has "proved to be effective in the support of many business activities not directly related to software engineering, but relying heavily on the

39

concept of process" [66]. This observation supports a common perception that the importance of process technology for software businesses as well as other businesses will increase in the future.

Essential components of the software process technology field are software process tools. Such tools are applied to provide process guidance, enable process automation, analyse processes, and to better understand processes in order to improve them [105]. A range of software process tools have been developed, for both commercial and academic purposes. Generally, the software process tools developed offer their own distinct functionality and focus on slightly different approaches.

A Process-centred Software Engineering Environment (PSEE) is the combination of several software process tools, with the aim of aiding developers in the software production process [105]. Both production and management technologies, as well as tools involved in a software process, are integrated in a process-centred environment, enabling the PSEE to implement, control and enhance the whole process [61]. Figure 2-8 shows how PSEEs and process technology can support the composite software process described in Section 2.3 and depicted in Figure 2-1.

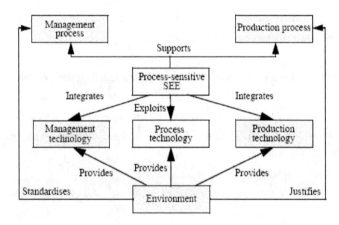

Figure 2-8. The Software Process Supported by Software Process Technology (from [61]).

PSEEs should generally provide support for process modelling, process enactment, and interaction with users and other computer systems. Process modelling support is typically provided by a process modelling language and tool, whereas process enactment support is provided in form of a process enactment engine.

2.6.1. Software Process Modelling

Software process modelling is the part of software process technology concerned with the definition of software processes in terms of for example tasks to be done and who is responsible. Process modelling today is often visual, resulting in a visual representation of the whole process. This makes it easier to follow and manage the process throughout its execution.

In his book, "The Art of Computer Programming", released in 1968, the author Donald Knuth amusingly compared processes to recipes for cooking [50]. He addressed the connection between these generally perceived distinct concepts by suggesting that a cook book is a collection of process descriptions, while the actual cooking activities using these descriptions are processes. In other words, the process description (recipe) is important for the outcome of the resulting product (meal). Osterweil built on Knuth's observed connection between process and product. He focused on the need for software processes to be modelled explicitly, in order to tailor them to the varying needs of complex and dynamic organisations [68]. For such explicit process modelling, he promoted the use of process modelling languages, and emphasised the importance of the resulting models being executable in order to enable ongoing process support throughout processes [68]. The long-term goal is to improve the organisations' products.

A process modelling language is typically referred to as a PML, and can be defined as an abstract formalism used to describe and represent processes in the form of process models [61]. In this context, process models are generally computer-internal descriptions of external processes [20]. Software process modelling may be defined as the creation of process models using PMLs or other modelling techniques. Dowson and Fernström describe two different categories of process modelling, namely descriptive and active modelling [27]. Descriptive modelling aims at describing processes and organisational behaviour, whereas the focus of active modelling is on producing process models that can be used to provide computer-based support for organisations and their processes. Process enactment systems represent the joining of these two process modelling ideas by making

41

use of process information from descriptive models together with process state information in active models.

Among several reasons for modelling processes, P. R. White identifies the most important ones as being [107]:

- To be more easily able to understand and reason about processes.
- To simplify communication about processes and help different views emerge.
- To be more easily able to analyse and improve processes.
- To enable enactment of the resulting process models in order to provide ongoing support and control throughout process execution.

A PML can be either formal, semi-formal or informal [20]. Formal PMLs are characterised by a formal syntax and semantics, and are therefore executable. Semi-formal PMLs generally have a formal syntax, but informal semantics, which means they are not executable. Neither are informal PMLs, of which examples are natural languages, such as English [20]. The focus in this research is on formal PMLs, as they provide a basis for enactment of process models.

Several different PMLs have emerged in the last few decades, ranging from purely textual representations on the one hand, to purely visual ones on the other. The different PMLs have been implemented and experimented with, but there have been no real efforts towards standardisation [20]. However, a general consensus on the most important process entities that should be included in a PML is beginning to emerge. Conradi and Liu argue that a basic PML must support at least the following six process elements [22]:

1. **Activities** – work that makes up processes.
2. **Artefacts** – products of processes.
3. **Roles** – responsibilities and rights for users filling the roles.
4. **Humans** – users that fill roles.
5. **Tools** – tools used in the completion of activities and tasks.
6. **Evolution support** – support for evolution of process models during enactment.

Currently existing PMLs have been built based on numerous different paradigms, leading to several attempts at identifying an appropriate classification of PML paradigms. Curtis et al. proposed a classification including five different paradigms; programming models, functional models, Petri net and state transition models, plan-based models, and quantitative models [26]. Lonchamp, on the other hand, identified six classes of PML paradigms, namely graphical, net-oriented, procedural, object-oriented, rule-based and multi-paradigms [54]. The four different classes of PML paradigms identified by Wang are as follows [105]:

1. **Programming language based** – the PML is a specialised programming language based on conventional programming languages.
2. **Rule-based** – the PML describes an activity with a rule, including pre-condition, action and post-condition, and the role and resources associated.
3. **Extended flow or network based** – the PML is a form of Petri net or state chart.
4. **Multi-paradigm based** – the PML is a combination of two or more paradigms.

An example of a programming language based PML is APPL/A, which focuses on the management of software objects in processes [86]. Examples of rule-based PMLs are Marvel's MSL and the PML used in Merlin. In MSL, each process step is encapsulated in a rule that has a name, multiple formal parameters, and the three optional constructs: condition, activity and effects [9]. The PML used in Merlin describes processes by rules that can be executed by both forward and backward chaining [70].

SPADE-1's SLANG and ProcessWeaver are examples of network-based PMLs. SLANG is a highly reflective language based on a high-level extension of Petri nets, where tokens are associated with values for example for activities and their states, and transitions are associated with actions [5]. The ProcessWeaver formalism describes the flow of tokens in a Petri net in a similar fashion [31]. EPOS' SPELL [21] and the CAGIS PML [105] are examples of multi-paradigm based PMLs, combining multiple PML paradigms.

Another important contribution to the PML domain is Swenson's Visual Process Language (VPL), which is a net-based visual notation that allows for collaboration during the planning process [87]. VPL is concise and easy to understand, and models a process as a composition of work plans at different levels [87]. Furthermore, VPL has been used as the basis for other PMLs, such as Serendipity's Extended Visual Planning Language (EVPL). EVPL extends VPL by preserving its simple work plans, and adding facilities for general process modelling to enhance the applicability of the language [40].

2.6.2. Software Process Enactment

The part of software process technology that deals with supporting real-time execution of processes is called software process enactment. It is typically associated with a process engine that drives the process by guiding process flow and assisting users in performing their allocated tasks.

Software process enactment support can be offered using a range of different approaches. Flexible guidance and total enforcement are two contrasting approaches, and various process support systems offer different degrees of guidance and enforcement. Dowson and Fernström identify four different levels of process support [27]:

1. **Passive guidance** – support is offered only on users' request.
2. **Active guidance** – users are guided, but not forced to perform actions.
3. **Process enforcement** – users are forced to act according to process specifications.
4. **Process automation** – the system completes actions without users' intervention.

Developed by Fraunhofer IESE, SPEARMINT is an example of a recent process tool that focuses on process guidance [16]. From a pre-defined process model, SPEARMINT allows for the generation of an electronic process guide (EPG), which provides interested users with information about the process and guidance on the steps to be performed [16].

Another recently developed process modelling tool is MILOS, which stands for Minimally Invasive Long-term Organizational Support [56]. The approach adopted in MILOS focuses on active process guidance, while at the same time providing some degree of automation of, for example, information flow [56].

Experience has generally shown that no single approach should be adopted in any situation [27]. Preferably, the approaches adopted should build upon knowledge about users and their requirements for the system being built, and the degree of enforcement, guidance and automation should be tailored to meet these requirements.

Software process enactment is usually implemented by events and triggers that cause changes in process state or process-related entities, such as artefacts. Additionally, means to guide users through the execution of processes and keep them aware of the state of the processes at any time is normally incorporated. A process enactment engine typically adopts the same paradigm as that employed by the process modelling language in the same environment. For example, if a PSEE uses a rule-based PML, it is likely to also employ a rule-based process engine.

Figure 2-9 shows an example of an enacted process model in the Serendipity-II environment [43]. The two main views show the enacted process model. It is highlighted with colours to show the state of the process and the stages that are currently being enacted. The bottom left-hand dialogue displays the enactment history of the currently enacted stage, and the one on the right shows other users and their enacted process stages.

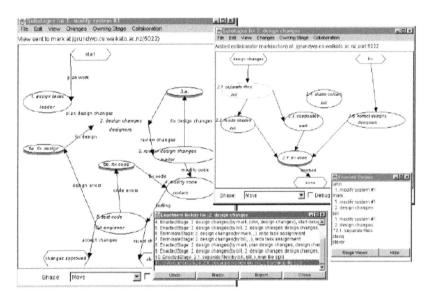

Figure 2-9. An Enacted Software Process in the Serendipity-II Environment (from [43]).

2.7. Related Disciplines

The purpose of this section is to highlight the scope of the work undertaken in this thesis. Totland and Conradi contend that the software engineering community is not the only one focusing on process technology [92]. They identify seven other fields that also

45

incorporate process technology as an important part of their framework. This section presents definitions and brief descriptions of these fields, followed by a comparison of all the fields. Totland and Conradi conclude their article by suggesting that because all the research areas have overlapping interests, their respective communities should be open for exchange of ideas and experiences [92].

2.7.1. Business Process Reengineering

Hammer and Champy define Business Process Reengineering (BPR) as "the fundamental rethinking and radical redesign of business processes to achieve dramatic improvements in critical, contemporary measures of performance, such as cost, quality, service, and speed" [44]. The main focus of BPR is clearly to improve an organisation's performance, and the authors further argue that information technology and human resource management are key factors in achieving such improved performance.

2.7.2. Computer Integrated Manufacturing

Computer Integrated Manufacturing (CIM) is defined by Rembold et al. as "the concept of a semi- or totally automated factory in which all processes leading to the manufacture of a product are integrated and controlled by computers" [74]. Simplified, CIM is the application of computer technology to the planning and control of manufacturing processes.

2.7.3. Concurrent Engineering

Concurrent Engineering (CE) was initially proposed as a potential means to minimize product development time, and Winner defines it as "a systematic approach to the integrated concurrent design of products and their related processes, including manufacturing and support" [108]. A main goal of CE is to make use of information technology to achieve the execution of many tasks in parallel, in order to reduce development time and cost.

2.7.4. Enterprise Modelling

According to Fox and Gruninger, Enterprise Modelling (EM) is the concept of building "a computational representation of the structure, activities, processes, information, resources, people, behaviour, goals and constraints of a business, government, or other enterprise" [32]. The main purpose of EM is further described as achieving model-driven enterprise design, analysis and operation.

2.7.5. Information Systems Engineering

Totland and Conradi have defined Information Systems Engineering (ISE) as the "application of a set of systematic engineering approaches to develop an information system", where an information system is "a system of computer components, software components, and human and organizational components that are developed, trained and assembled to fulfil the information processing requirements of a problem" [92]. ISE focuses on the development of a working system as a whole, including computers, software and humans.

2.7.6. Organisational Design

According to Navarro, Organisational Design (OD) "views an organisation as an open system of interrelated 'elements' that transforms various inputs into desired outputs", and "explicitly addresses both technical and social aspects of work organizations, as well as the interrelation between them in the performance of the work system as a whole"[63]. The focus of OD is on improving organisational performance by applying computer technology to the design of organisations and work processes. Organisational design is influenced by a number of factors [33], the most common being the organisation's formal structure, the resource access structure and procedures for training personnel [17].

2.7.7. Workflow Management

Workflow Management (WM) can be described as the control of workflow, which is defined by the WfMC as "the automation of a business process, in whole or part, during which documents, information or tasks are passed from one participant to another for action, according to a set of procedural rules" [109]. Hence, a major emphasis of the WM area is the management of the flow of documents, information and tasks.

2.7.8. A Comparison of the Application of Process Technology

Based on the work by Totland and Conradi [92] and on the most essential process elements identified in Section 2.6.1, Table 2-1 presents a taxonomy and the terminology for comparing the software engineering field's use of process technology with that of the related disciplines. Domains are the areas the discipline focuses on, objectives are the primary purposes for applying process technology within those areas, and process elements are the most commonly used process modelling elements within the areas.

Table 2-1. Taxonomy and Terminology for Comparing Related Disciplines (after [92]).

DOMAINS	Software production
	Manufacturing
	Information services
	Physical services (health services etc.)
OBJECTIVES	Comprehension – understanding real-world processes
	Collaboration – improving human-human interaction
	Evolution – evolving processes
	Analysis & Design – analysing & designing processes and their activities
	Coordination – coordinating human actors and activities
	Enactment – computer-driven execution of processes
	Evaluation – evaluating, monitoring, and gathering data about processes
PROCESS ELEMENTS USED	Activities – work segments that together make up a process
	Artefacts – products of a process
	Roles – responsibilities and rights for users filling the roles
	Humans – users that fill roles
	Tools – tools used in the completion of activities and tasks
	Evolution – support for evolution of the process model during enactment
	Business Rules – rules and guidelines for work on activities and tasks

Table 2-2 shows a comparison of all the related fields, including the software engineering field, according to the taxonomy presented in Table 2-1. The focus of the software engineering field, referred to as SE in Table 2-2, is on the software production domain. Basically, all the objectives and process elements included in the taxonomy are relevant for this field, as it is one of the main fields in which process technology is used. The only one not relevant for the software engineering field is the 'business rules' process element.

Table 2-2. A Comparison of the Related Disciplines (after [92]).

DISCIPLINES*		SE	BPR	CIM	CE	EM	ISE	OD	WM
DOMAINS	Software production	√	√		√	√		√	
	Manufacturing		√	√	√	√		√	
	Information services		√		√	√	√	√	√
	Physical services		√		√	√		√	
OBJECTIVES	Comprehension	√	√	√	√	√	√	√	√
	Collaboration	√	√		√	√	√	√	√
	Evolution	√	√	√	√	√	√	√	√
	Analysis & Design	√	√	√	√		√	√	√
	Enactment	√		√			√		√
	Evaluation	√		√					
PROCESS ELEMENTS USED	Activities	√	√	√	√	√	√	√	√
	Artefacts	√		√			√	√	√
	Roles	√	√		√	√	√	√	√
	Humans	√			√	√		√	√
	Tools	√		√				√	
	Evolution	√							
	Business Rules		√			√	√	√	√

* SE - Software Engineering, BPR - Business Process Reengineering, CIM - Computer Integrated Manufacturing, CE - Concurrent Engineering, EM - Enterprise Modelling, ISE - Information Systems Engineering, OD - Organisational Design, WM - Workflow Management

CIM concentrates on manufacturing, whereas the focus of ISE and WM is on the information systems domain. The remaining disciplines, BPR, CE, EM and OD, apply to all the domains included in the taxonomy, namely software production, manufacturing, information services, and physical services.

When it comes to objectives, comprehension and evolution are important reasons for applying process technology in all the related fields. Furthermore, collaboration is a primary objective of all except the CIM field, and analysis and design are important purposes for all except the EM field. Aside from the software engineering field, only WM, ISE and CIM are concerned with process enactment, and CIM is the only other field that uses process technology for evaluation purposes.

The most commonly used process elements are activities, which are used by all the related fields. Roles are also common, in that they are utilised by all the fields apart from the CIM field. The remaining process elements used by the software engineering field are artefacts, humans, tools and evolution. Evolution is the only one not employed by any of the related fields. Artefacts, however, are utilised by CIM, ISE, OD and WM. Furthermore, the CE, EM, OD and WM fields make use of humans, and the CIM and OD disciplines use tools. Business rules, which are the only process elements not employed by software engineering, are utilised by the BPR, EM, ISE, OD and WM fields.

2.8. Workflow Management Systems

A Workflow Management System (WFMS) is a system that "defines, creates and manages the execution of workflows through the use of software, running on one or more workflow engines, which is able to interpret the process definition, interact with workflow participants and, where required, invoke the use of IT tools and applications" [109]. WFMSs are closely related to PSEEs, and the two terms are often used interchangeably. However, the main focus of WFMSs is on the flow of documents, information and tasks in a general office environment, whereas PSEEs has more of a focus towards assisting the execution of software production processes by, for example, simple task automation and software tool integration. The following paragraphs examine some existing WFMSs and their essential features.

The wOrlds environment is a WFMS that makes use of 'obligations' to handle flexible, dynamic changes to workflows [15]. An obligation provides a method to generate a request from one user to other users, and workflows in wOrlds are composed of networks of such obligations [15].

Another environment, Oz, presents an approach to integrating individual synchronized groupware activities into modelled workflow processes [10]. An objective of the Oz

WFMS is to show how workflow and groupware activities can be integrated in order to exploit the advantages of both worlds [10].

InConcert is a commercial database-dependent and process-oriented WFMS that supports hierarchical decomposition of tasks [10]. It helps individuals keep track of their work, and provides easy access to documents that are to be worked on. Task automation is supported by assigning a task to a computer program instead of a human being. Another important feature of InConcert is that it focuses on the modelling of work, rather than documents, by providing a shared information space containing all documents needed in a process [10].

According to Van der Aalst, the current situation regarding workflow management software is similar to that of database management software in the early 1970s; a lack of standardised methods led to the use of ad-hoc approaches for the development of Database Management Systems (DBMSs) [94]. However, as standards such as the Relational Data Model [19] and the Entity-Relationship Model [18] evolved, these served as a common basis for the development of DBMSs, leading to a boost in their use. The WfMC is currently working on standards for workflow management software, but they have yet to be widely recognised and used, and many organisations are reluctant to use currently available WFMSs. However, as appropriate standards evolve, we might see as a result a similar boost in the use of workflow management software as that of DBMSs following the 1970s.

2.9. The Pounamu Meta-CASE Environment

Pounamu is a new visual software engineering tool under development by Auckland UniServices Limited at the University of Auckland [110]. The current version of Pounamu, Version 1.0, allows users to model software engineering projects, as well as to create visual software engineering tools in which Pounamu itself is a core part. Hence, Pounamu can be used both as a CASE (Computer Aided Software Engineering [36]) tool and a meta-CASE tool.

Two important features differentiate Pounamu from similar tools. These two features are

51

dynamic generation of shapes and connectors for use in modelling, and dynamic generation of event handlers to deal with modelling events [110].

Pounamu consists of a number of sub-tools. However, the two main tools are the tool creator and the project modeller, which can be used alternately. A screen dump of the Pounamu user interface on start-up is shown in Figure 2-10, with the main components marked in red. The main components and their purpose are:

1. **The main menu bar** – contains access to top-level functionality.
2. **The tree panel** – manages opened projects and their entities.
3. **The main showing area** – displays selected projects and entities.
4. **The property area** – shows the properties of selected entities.
5. **The information area** – displays modelling-related information.

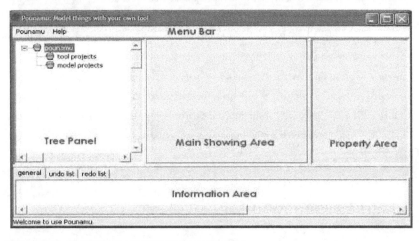

Figure 2-10. The Pounamu User Interface on Start-Up.

The following two sections give a further description of Pounamu's two major tools, namely the tool creator and the project modeller.

2.9.1. The Tool Creator

The tool creator is a meta-CASE tool that can be used to create new software tools [110].

It consists of several sub-tools, which provide different functionality. The sub-tools and the functionality they offer are as follows:

1. **The shape creator** – creation of custom shapes.
2. **The connector creator** – creation of custom connectors.
3. **The handler definer** – definition of custom event handlers.
4. **The meta-model definer** – definition of custom meta-models composed of entities and associations.
5. **The view-type definer** – definition of different view-types by mapping available shapes to entities, and connectors to associations.

When a new tool project is opened in Pounamu, the five sub-tools are added to the project by default. The user can then use these tools alternately to create a new modelling tool. Access to the different tools can be obtained in the tree panel or via tabs at the bottom of the main showing area. User-defined entities within each sub-tool can be accessed via the tree panel or via the tabs at the top of the main showing area.

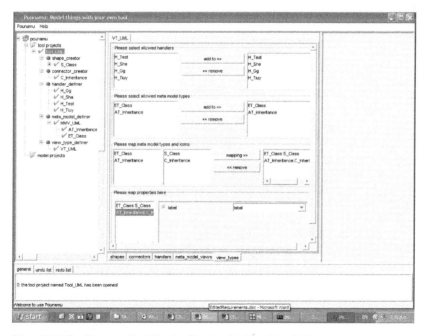

Figure 2-11. The Pounamu Tool Creator Showing the View-Type Definer.

53

Pounamu is dynamic, meaning that all the other parts of the tool will be updated according to any changes made to any one part of the tool. For example, if a new property is added to a shape in the shape creator, this shape is updated in the view-type definer accordingly. The screen shot in Figure 2-11 shows a view-type definer in the Pounamu tool creator.

2.9.2. The Project Modeller

The project modeller, depicted in Figure 2-12, is a CASE tool that can be used to model software projects [110]. A model project is associated with a user-defined tool, and used to model a real-world case. When the user opens a new model project, he/she chooses the tool he/she wants the new model project to be associated with. The new model project automatically contains each view-type that exists in its corresponding tool.

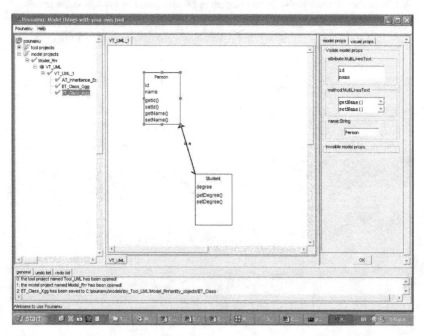

Figure 2-12. The Pounamu Project Modeller.

Both multiple view-types and multiple views are supported in Pounamu [110]. A tool can have multiple view-types that are, as mentioned, automatically contained in a model project associated with the tool. In a model project, multiple views can be created within each view-type inherited from the corresponding tool. This is very advantageous when a diagram or a model grows too big to fit in one view, as it can easily be split into several smaller views within the same model project.

Another special feature of Pounamu is the fact that an entity, such as a shape or a connector, can appear in several views of different view-types. It always appears with the same icon in views within the same view-type, but can map to different icons in different view-types. All the icons mapping to the same entity can be thought of as sharing the entity, and Pounamu facilitates coordination among these icons [110].

2.10. Web Services Technology

A Web Service is essentially a software program that supports interaction between other software programs that reside on different machines in dispersed locations [101]. It can be seen as a middleware that facilitates application-application or application-user joins by providing a bridging mechanism between them using Web protocols. This is achieved by means of remotely accessible services, to which Web Services owe their name; they provide services to their users, being it a human user or another application. Examples of services that they can provide are everything from company database access to functionality for travel bookings. A more detailed description of Web Services and their advantages is included in Section 5.3.

2.11. Summary

Software engineering, software process and software process technology are all identified as research areas that are significant for the work done in this project. Workflow Management Systems and the Workflow Management field, along with various other disciplines, are also closely related, and their study widens the understanding of the scope of the work undertaken. Furthermore, the Pounamu meta-CASE tool provides flexible and extensible functionality for creation of visual tools as well as visual modelling using the tools that have been created. Finally, Web Services technology can be used as a means to facilitate communication between applications and achieve service-orientation.

Chapter 3 - Requirements

3.1. Introduction

Object-Oriented Analysis (OOA) involves capturing the high-level requirements of a system that is to be built. This chapter provides such an analysis of the requirements underlying the IMÅL process support environment. General and specific high-level functional requirements are proposed, as well as non-functional requirements. Further investigation of the identified requirements is then presented in the form of UML (Unified Modelling Language [72]) use cases and OOA diagrams in order to enhance comprehension and aid in formalising the design.

3.2. Functional Requirements

Functional requirements describe the behaviours of a system in terms of the functionalities and services it offers to its users [55]. This section firstly describes general functional requirements for the IMÅL process tool. Thereafter, the focus is on the functional requirements specific to the prototype implementation realised in this project, as time constraints made it necessary to narrow the general specification.

3.2.1. General High-Level Functional Requirements

As described in Section 1.2, an important aim of this thesis work was to investigate some novel concepts in the software process technology area. This objective was to be realised through the development of a decentralised, service-oriented process management and support environment. Among the most important features that should be offered by the environment were a visual programming language and tool, a heterogeneous and distributed process enactment engine built up of several smaller Web Services with different responsibilities, and seamless integration with third-party tools.

Presented in the following are general high-level functional requirements identified for the IMÅL process management and support tool, including functionality for both process modelling and enactment. The requirements are listed below, and then explored individually in the subsequent sections.

The IMÅL process tool should:

- Provide a visual process modelling formalism and tool.
- Provide a way for the user to interact with the system.
- Provide a distributed and service-oriented process enactment engine.
- Provide communication, coordination and collaboration facilities for both process modelling and enactment.
- Facilitate seamless integration with third-party tools.
- Support evolution of processes and process engine.

3.2.1.1. A Visual Process Modelling Formalism and Tool

The system should provide a visual formalism that can be used to model processes, as well as serve as a basis for process enactment. Such formalism is normally provided through a visual PML in a process modelling tool. There are several requirements for a PML, in order for it to provide the appropriate support. Listed and elaborated on below are the most important requirements for the PML in IMÅL.

The PML should:

- Be formal enough to enable enactment of the process model, yet remain simple and intuitive.
- Support abstract modelling of work processes.

58

- Allow for modelling of at least the six basic process elements listed in Section 2.6.1.
- Allow for modelling of associations between tasks and processing instances.
- Provide means to describe human-tool integration.
- Support analysis and evolution of processes and process models.

The importance of the PML being formal was driven by the desire to enable enactment of the resulting process model. A formal PML opens for process interpretation and execution by a process engine, as opposed to an informal PML, in which the informality makes it nearly impossible to use a standard process engine to interpret and execute the model. However, although formal, the process modelling language should be relatively simple, intuitive and easy to learn and understand.

Supporting abstract modelling of work processes was another essential requirement for the PML. An example of abstraction is the distinction between a role and a human user. When modelling processes, a role rather than a specific human user is associated with each activity. Human users are consequently modelled to fill specific roles. Distinguishing between human users and the roles they fill to support abstraction adds flexibility and semantic to the PML.

In order to optimise the use of the multiple processing instances that make up the process engine, the PML should allow the user to allocate processing of tasks to appropriate processing instances. This gives the user flexibility to manually define which instances are responsible for processing certain tasks. If no processing instance is specified, a default instance should be used.

The PML should allow for modelling of at least the following elements: activities, artefacts, roles, humans, tools and evolution support. Including these six basic process elements in a PML is commonly regarded as a minimum requirement [22]. However, a real-world process can be very large and include numerous process elements of importance. The PML should therefore preferably also include support for modelling of additional process-related elements, such as projects, workspaces, and work coordination and collaboration [22].

Appropriate means to describe human-tool integration should also be provided by the PML. Human-tool integration refers to the support for interaction between users and other tools or applications external to the system itself. A way of doing this is to have the system automatically present the user with the tools they need to perform their tasks at any time. The modelling of human-tool integration is essential in order to enable such tool support. It is worth noting that tool-tool integration – supporting the interaction between the system and other tools – might also be of importance. Modelling of tool-tool integration enhances the system by enabling a third-party computer program to automatically process specific tasks that need no user intervention.

Finally, the PML should support the analysis and evolution of processes and process models. As real-world processes are subject to change, the analysis and reflection of such changes should be supported by the PML. Such support is desired in order to repeatedly improve and advance process models and their execution.

3.2.1.2. User Interaction with the System

User interaction with the system is usually provided via the tool used for process modelling, which means that the process can be modelled and enacted using the same tool. The main benefit of offering all user interaction within the process modelling tool is that the user only needs to learn and work with one tool. Additionally, it makes sense to provide the user with enactment functionality via the process modelling tool, seeing that this is where a visual representation of the process definition resides. On the basis of this, IMÅL should provide both process modelling and enactment functionality within the same tool.

3.2.1.3. A Distributed and Service-Oriented Process Enactment Engine

The process engine is an essential component at the heart of the IMÅL process support system. Fundamental requirements for the process engine include providing a way of interpreting process models, as well as providing functionality for execution of the interpreted models. The following lists and describes detailed requirements for IMÅL's process engine.

60

The process engine should:

- Be a distributed and heterogeneous collection of processing instances.
- Allow for enactment of a process model and automation of parts of it.
- Store information about process state.
- Store information about process models.
- Support the ability to handle process-related events.
- Support the generation of to-do lists for users.

A requirement for the IMÅL process engine, and a novelty of this work, was that it should be implemented as a distributed and heterogeneous collection of multiple processing instances. The reason behind doing this was to better distribute the process engine, and to efficiently enable the delegation of certain processing tasks to the instances responsible for the respective tasks. As described in Section 2.6.1, several different process modelling paradigms exist, such as rule-based and Petri net based. By building a process engine as a collection of multiple processing instances, all these different paradigms can be supported by one engine, which means that the advantages of each approach can be utilised. For example, one processing instance could be responsible for simple event-based processing, another for rule-based processing and a third for Petri net based processing. Also, with such a heterogeneous environment, the failure of one processing instance or site does not cause the whole engine to fail. The other processing instances can keep processing their allocated tasks while the failed instance is repaired, and the core process execution need therefore not be interrupted.

Crucial for allowing computerised enactment of a process is providing a way of interpreting the corresponding process model [61]. The formal process model should therefore be stored in a format that the engine can interpret and translate into a representation internal to the engine. Once the engine has such a representation, it can support process execution by working with this internal representation of the process model.

Automation is another important aspect when it comes to process enactment [61]. IMÅL's process engine should offer automation of process flow and of tasks that can be completed without user intervention. Automation of process flow means that when one

61

activity is finished, the process engine automatically changes its state, and determines the next step in the process. Task automation implies that tasks are completed automatically in the background by the process engine itself, or other tools triggered by the process engine, without any involvement from the user. An example of task automation is automatic notification about process progress to all users involved. Such automatic functionality is closely related to the modelling and execution of tool-tool integration, and contributes to enhance the efficiency of processes and their execution.

Another important requirement for the process engine was that it should store information about process models and process state. Information that needs to be stored about process models includes descriptions of process elements, such as activities, artefacts, roles, humans and tools. The process state information, on the other hand, refers to the state of each activity in a process. The process engine needs to store this information in order to aid process execution, assist the user in completing work, and keep track of what is going on in a process at any given time.

Furthermore, the handling of process-related events is essential in enabling computerised execution of process models [15]. Process-related events include events such as process state changes, artefact updates, and events in work tools. Both the user and the engine itself should be able to trigger these events. Each time such an event occurs, it should be forwarded to the part of the process engine that deals with that type of event. The appropriate portion of the process engine should then determine and initiate the actions to be taken.

Finally, the process engine should support the generation of to-do lists for users involved in a process. When a process model is enacted, a user involved in the process should be able to, by supplying a valid username and password, obtain a list of all the activities he/she is responsible for in this process. This should include essential information about the activities, including their current states. An overall view of the state and progress of the process should also be provided. For increased process support, the to-do list should also contain functionality for user interaction with the system in a similar fashion to the process modelling tool. Hence, the user should be able to enact processes directly from within the to-do list.

3.2.1.4. Communication, Coordination and Collaboration Facilities

Providing facilities to support cooperative work is increasingly important in software systems today. The areas of Computer Supported Cooperative Work (CSCW [77]) and Groupware [28] are receiving continuous attention as a result of new technology leading to computers now housing a single framework for interactions that were previously supported using different technologies over different networks [45]. IMÅL should provide facilities to coordinate teams that are dispersed at different locations according to the detailed requirements listed and described in the following.

The IMÅL environment should:

- Support work coordination.
- Support collective work.
- Allow for synchronous and asynchronous collaborative editing of process models.
- Support synchronous and asynchronous user-user communication.
- Provide awareness about process progress.

As with any process support system, an essential feature of IMÅL should be the support for coordination of work in a distributed team. The coordination of distributed work is essential when it comes to increasing the efficiency of tasks and processes. Seeing that IMÅL aims to provide efficient process management and support, work coordination issues should therefore be carefully addressed. Users should be able to access important process-related information and perform work regardless of their location. As described previously, automation of process flow and tasks would also contribute to enhance the coordination of group members involved in a process.

Additionally, the system should provide support for collective work, both when modelling processes and when working on activities in enacted processes. This is important in order for the system to conform to real-world process scenarios where people often work collectively to solve problems and complete tasks.

Related to the support of collective work is the enabling of collaborative editing of process models. Both synchronous editing, where changes one user makes are updated immediately in collaborative users' model, and asynchronous editing, where changes are stored and updated at a later time, should be supported. Synchronous and asynchronous user-user communication should also be allowed, both when editing process models and when working on an enacted process.

Finally, the system should continually keep the users up to date on process progress. Users involved in a process should be made aware of what the state of the process is at any time, who is working on what, and what the next process step is. The reason this is important is that by making users aware of what is going on in the process at any time, efficiency is increased and bottlenecks largely avoided.

3.2.1.5. Seamless Integration with Third-Party Tools

IMÅL should provide good integration of process modelling tools, work tools and communication and editing tools. In particular, the focus in this project was on the integration of Web Services applications. This focus was aimed at investigating a fully decentralised approach to process modelling, management and support. Also, building the system as a collection of Web Services components to support a component-based architecture increases the stability of the system, and reduces the impact of possible failures in single components. For example, if one service is out of order, the system can still go ahead and use the remaining services without problems, instead of the whole system coming to a standstill because an error has occurred in one small part of it. Another advantage of the component-based approach is that it provides openings for other components and tools to be conveniently incorporated into the system in the future.

The reason behind providing such tool and service integration is that it enhances the users' perception of working in one large environment that offers all the functionality they need to do their work, rather than using numerous separate tools with different specific functionality in order to achieve their goals. Providing good integration with external tools is also essential in order to assist software engineers in performing the actual process work.

An example of good integration of external tools and user assistance is when a user enacts a part of the process, and the program automatically opens the document associated with this stage and displays to the user an explanation of the work he/she is required to do. When the user finishes working on the document, the system detects this and prompts the user for information in order to progress the process further.

3.2.1.6. Support for Evolution of Processes and Process Engine

Dynamic change of processes should be supported. This includes functionality to update a process model while it is being enacted. Supporting dynamic change leads to more efficient processes and process models. Furthermore, the system as a whole should evolve along with improved processes and process models as well as changes occurring in the surrounding environment.

3.2.2. Prototype-Specific High-Level Functional Requirements

Because of the time constraints associated with this thesis project, the high-level requirements presented in the previous section were prioritised for the development of a prototype system. Some of the requirements were awarded special attention, whereas others received little attention or none at all. This section describes the prioritisation of the requirements for the prototype development.

The main focus for the prototype was on the implementation of a service-oriented, distributed and heterogeneous process enactment engine to provide interpretation and execution of process models. All the requirements outlined for the process enactment engine in the previous section are therefore highly relevant. The process engine should be composed of a collection of smaller Web Services, including multiple processing instances adopting different processing paradigms. A simple event-based processing instance supporting simple process execution was first to be developed, followed by a rule-based approach and others if time permitted.

An important functionality that should be provided by the process engine was the generation of to-do lists for users. Some requirements were identified for this functionality. Firstly, the to-do lists should be accessible via the Internet. Users should be able to access the login page for the to-do list service by pointing any Web browser to the address of the service. Username and password provided by the user should be checked against user information stored in a database, and access should be granted only if the username and password conforms to that of a user in the database. When logged in, the user should be able to view up-to-date information about enacted processes.

Further, a process modelling formalism and tool were needed in order to model processes. As the main focus was on the actual engine, a deliberate decision was made to spend less time and effort on process modelling support. This could be achieved, as the Pounamu visual modelling tool, which allowed for a basic process modelling language and tool to be created relatively rapidly, was available in-house. Pounamu was also chosen as the front end of the system, providing the user interface for both modelling and enactment. A more detailed rationale for why Pounamu was chosen is included in Section 5.2.

Building on the explanations above, and on the general requirements for the IMÅL process support tool presented in the previous section, the following lists the prototype-specific system requirements in order of prioritisation from the top down.

- **Provide a distributed and service-oriented process enactment engine that:**
  - Constitutes a distributed and heterogeneous collection of processing instances.
  - Allows for enactment of a process model and automation of parts of it.
  - Stores information about process state.
  - Stores information about process models.
  - Supports the ability to handle process-related events.
  - Supports the generation of to-do lists for users.
- **Use Pounamu to provide a process modelling formalism and tool that:**
  - Is formal enough to enable enactment of the process model, yet remains simple and intuitive.
  - Supports abstract modelling of work processes.
  - Allows for modelling of at least the six basic process elements.
  - Provides means to describe human-tool integration.
- **Provide a way for the user to interact with the system via the Pounamu user interface by:**
  - Updating the user interface according to process-related events.
  - Updating the process model according to process-related events using colours.

3.3. Non-Functional Requirements

Non-functional requirements are requirements that deal with the constraints and qualities of a system [55]. Several non-functional requirements were identified for the IMÅL prototype process system. They are presented and explained in the following.

3.3.1. Usability

Users that have experience with visual modelling tools should find IMÅL relatively easy to learn and use. However, some experience with and knowledge about processes and their management is required for optimal use of the system. The PML should be reasonably intuitive, although some degree of syntax must be learnt in order to model processes efficiently. Easily understandable process enactment functionality should be offered to users in a straightforward manner, with little room for enactment errors.

3.3.2. Extensibility

The addition of new features to the system should be conveniently achieved, and the system should tolerate having unspecified as well as specified extensions added to it in the future.

3.3.3. Reusability

The IMÅL process engine component as a whole should be reusable in any circumstance where the process definition is available in the specific format the process engine takes. In other words, the process engine should be able to run a process model defined using any process modelling tool, as long as the definition is translated to the appropriate format before it is forwarded to the engine.

3.3.4. Pluggability

The process engine component should be added to Pounamu in a component-based

67

fashion, implying that the facility can be dynamically plugged into Pounamu as the user desires, rather than being built in as a core part of the system [41].

3.3.5. Middleware

IMÅL's process engine should be built as a collection of Web Services, using Web protocols to facilitate communication between Pounamu and the process engine, and between the services that make up the process engine.

3.3.6. Hardware

A computer with a moderate processor should be sufficient for the running and use of the IMÅL system, including Pounamu. The services that make up the process engine should also be able to reside on moderate-sized computers, with moderate processors and sufficient memory. For optimisation purposes, the services should be dispersed on several computers at different locations. If necessary, however, it should be possible to deploy all the services on a single server.

3.4. Overall System Diagram

Based on the requirements outlined in the previous sections, Figure 3-1 presents the main components in the IMÅL system and how the integration of the Pounamu software tool was achieved. The project manager uses modelling facilities provided by the existing Pounamu tool to model a process. When he/she wants to enable enactment of the process model, he/she starts up the process engine by choosing the 'plug in process engine' option from the Pounamu project's menu bar (1). Starting the process engine service from Pounamu involves sending it the process definition, from which the service creates its own local representation of the process. Appropriate updates to the user interface and the process model are then applied to Pounamu (2), making the process ready for enactment. The user is now able to run the process by enacting, pausing, resuming and finishing stages in the process. Each such enactment event is automatically forwarded to the process engine service (3), where the event is processed with help from other external services. After each enactment event that occurs, the user interface and process model in Pounamu is updated according to the results of processing the event (4).

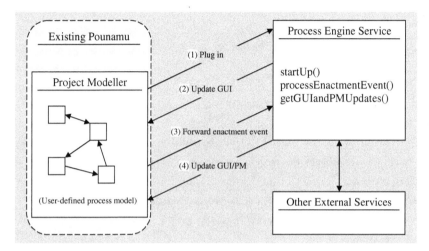

Figure 3-1. Overall System Diagram.

3.5. Use Cases

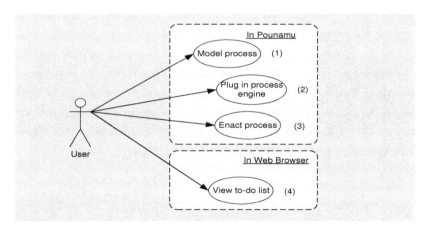

Figure 3-2. Use Case Diagram Capturing Interactions Between Users and the System.

The use case diagram shown in Figure 3-2 captures all the possible interactions between an external user and IMÅL in four use cases. In the following, each use case is described in detail including screen shots depicting the use cases in action in the prototype system.

3.5.1. Use Case 1: Model Process

Name: Model process.

Description: Model a process using Pounamu.

Actors: Process modellers.

Precondition: A process modelling tool must be available.

Result: An executable process model.

Flow of events:

    1. The user opens a model project associated with the process modelling tool.

    2. The user models a process by adding shapes and connectors to views in the model project.

    3. The user chooses 'save project' from the project menu.

    4. Pounamu saves the model project into an XML file.

Figure 3-3 shows an example screen shot of Use Case 1. A start-stage, a stop-stage, three base-stages, and five flows, have already been added to the view, and the user is about to add a role-actor association. Once the user has finished adding role-actor associations, the view contains a complete process model, and the process engine can be plugged in.

The complete process model resulting from adding the desired role-actor associations is depicted in Figure 3-4. Named 'Model_UpdateSoftware', it is a model of a simple process that updates an existing software program by adding new functionality to it. This scenario will be referred to throughout the rest of this thesis report in order to demonstrate the functionality provided by the IMÅL prototype tool and how it can be used.

70

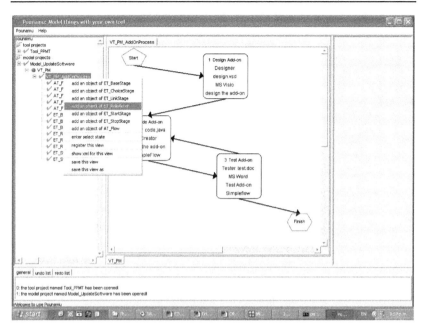

Figure 3-3. Use Case 1: Model Process.

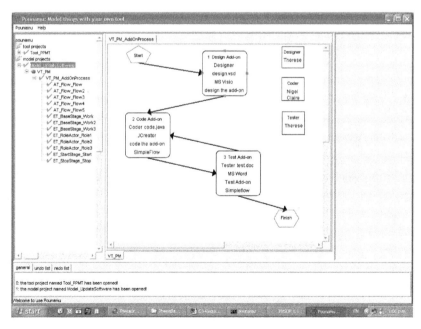

Figure 3-4. The Complete Process Model for a Software Update Scenario.

71

3.5.2. Use Case 2: Plug in Process Engine

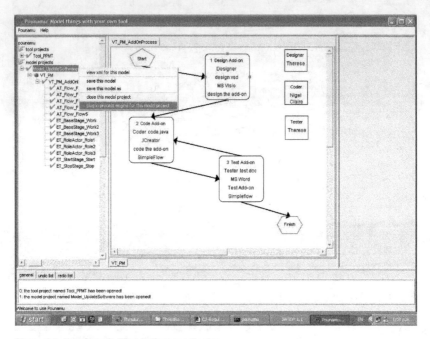

Figure 3-5. Use Case 2: Plug in Process Engine.

Name: Plug in process engine.

Description: Plug in a process engine component to a running Pounamu.

Actors: All certified users.

Precondition: A process modelling project to which process engine functionality can be plugged in must be available.

Result: The process engine component is plugged in and initialised and the process model is ready to be enacted.

Flow of events:

    1. The user chooses 'plug in process engine for this model project' from the project menu.

    2. Pounamu initialises its process engine component.

        2.1. Pounamu's process engine component starts up the process engine service and feeds to it the process definition.

1.1.2. The process engine service starts up and initialises other external services it uses to support enactment.

3. Pounamu's process engine component calls the refresher service to obtain necessary updates to the user interface, and applies them in Pounamu.

An example screen shot of Use Case 2 is shown in Figure 3-5. The user chooses 'plug in process engine for this model project' from the model project drop-down menu, and the process engine is plugged into the corresponding model project. When this is done, the process is ready to be enacted.

3.5.3. Use Case 3: Enact Process

Name: Enact process.

Description: Enact a process by starting, pausing, resuming or finishing work on a stage.

Actor: All certified users.

Precondition: The process engine component must be plugged in.

Result: Ongoing process support during work on stages.

Flow of events:

1. The user chooses an enactment event from a stage's drop-down menu.

2. Pounamu's process engine component forwards the enactment event to the process engine service.

2.1. The process engine service delegates the enactment event to the appropriate external service for processing.

1.1.2. The external service processes the enactment event, and records in the refresher service updates to the user interface and process model accordingly.

3. Pounamu's process engine component calls the refresher service via the process engine service to obtain necessary updates to the user interface and process model, and applies them in Pounamu.

Figure 3-6 displays an example of Use Case 3. The user has started the process from the start-stage's drop-down menu, and is about to enact the first stage. Once the process has been enacted, the user can make use of the to-do list service for ongoing information about the process.

73

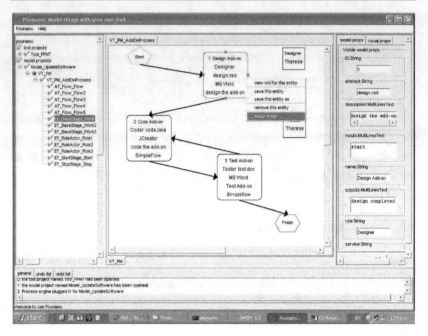

Figure 3-6. Use Case 3: Enact Process.

3.5.4. Use Case 4: View To-Do List

Name: View to-do list.

Description: View to-do lists by logging on to a Web-based to-do list service.

Actor: All certified users.

Precondition: At least one process model must be enacted.

Result: A to-do list is displayed in the Web browser.

Flow of events:

1. The user points a Web browser to the address of the to-do list service.

2. The service displays empty username and password fields as well as login and reset buttons.

3. The user supplies a username and password and hits the login button. (If the user hits the reset button, go back to 2.)

4. The service checks username and password against user records in a database. If

username and password are correct, the service displays a welcome message and a view to-do list button. If the login attempt is incorrect, go to 7.

5. The user hits the view to-do list button.

6. The service retrieves process and enactment information stored in databases, and displays a list of to-do items specific to the user logged in as well as a list with overall process information. These lists are automatically refreshed every minute.

7. The service displays a message indicating what was wrong with the login attempt. Go back to 2.

The login screen for Use Case 4 is depicted in Figure 3-7. The user enters username and password in the text fields shown, and hits the login button to login. If the username and password are valid, the user can click on a view to-do list button to view information about his/her responsibilities in the process as well as overall process information. Example screen shots of such to-do lists are included in Section 5.3.5 and Section 7.7.

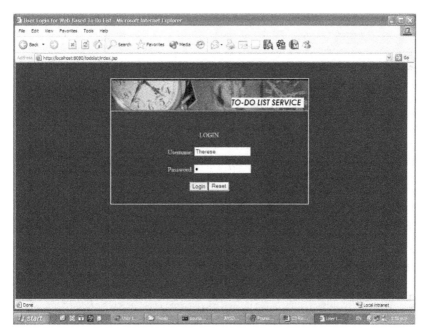

Figure 3-7. Use Case 4: View To-Do List.

75

3.6. OOA Class Diagrams

Based on the requirements identified for the system, three OOA class diagrams were assembled. The first diagram, shown in Figure 3-8, shows the overall composition of the Pounamu process engine component. Figure 3-9 and Figure 3-10 depicts the overall structure of the process engine, with Figure 3-9 showing the most essential services that make up the process engine and Figure 3-10 showing less essential but very useful services. Brief explanations of the classes included in the diagrams are given in the following two sections.

3.6.1. The Pounamu Process Engine Component

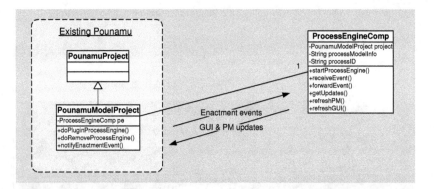

Figure 3-8. OOA Class Diagram for the Pounamu Process Engine Component.

PounamuModelProject: This class already existed as an essential part of the Pounamu software tool's project modeller. It is a sub-class of the more general PounamuProject class. However, some functionality had to be added to PounamuModelProject in order to support the plugging in of the process engine and the enactment of a process model. Most of the user interaction with the IMÅL system takes place within this class, as it allows a user to model processes, plug in the process engine, and enact process models.

ProcessEngineComp: Class ProcessEngineComp is the main class of Pounamu's process engine component. It contains a pointer to the PounamuModelProject that it is associated

76

with, process model information, and methods to start the process engine service and receive and forward enactment events. Additionally, it is responsible for obtaining updates to the user interface and process model resulting from enactment, and applying them to the model project.

3.6.2. The Process Engine Service

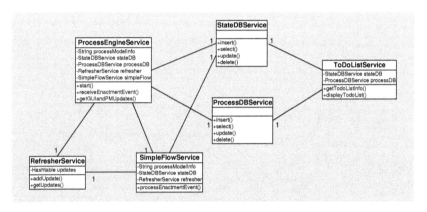

Figure 3-9. OOA Class Diagram for the Most Essential Services in the Process Engine Service.

ProcessEngineService: Class ProcessEngineService constitutes the heart of the process engine. This class acts as the overall coordinator of process enactment, making use of several different external services to do so.

StateDBService: The StateDBService class offers functionality to store and update process state information in a database.

ProcessDBService: The ProcessDBService class offers functionality to store process-related information contained in process models, such as id, name and description.

SimpleFlowService: SimpleFlowService is a class that class ProcessEngineService makes use of in order to support simple event-based process flow. Its main task is to process

77

received enactment events, including updating the StateDBService and the RefresherService classes.

RefresherService: The RefresherService class is used to store and retrieve updates to the user interface and process model in Pounamu, according to the result of processing enactment events.

ToDoListService: Class ToDoListService provides functionality to retrieve state and process information using the StateDBService and the ProcessDBService classes respectively, as well as functionality to display the retrieved information neatly in a Web browser.

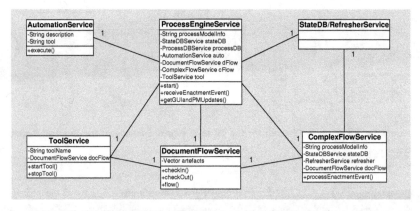

Figure 3-10. OOA Class Diagram for Additional Services in the Process Engine Service.

ComplexFlowService: Class ProcessEngineService uses class ComplexFlowService to support more complex flow that involves several variables and components. Like class SimpleFlowService, it processes enactment events, and updates class StateDBService and RefresherService according to the results. Several complex flow instances supporting different processing paradigms, such as rule-based and network-based, could be included.

DocumentFlowService: Functionality to enable and coordinate the flow of documents according to process enactment is contained in class DocumentFlowService.

ToolService: Class ToolService provides functionality to automatically initiate third-party tools that are needed in order to perform work on enacted processes.

AutomationService: Class AutomationService is responsible for automatically performing certain tasks or parts of processes that need no user intervention.

3.7. Summary

Several requirements were identified for the IMÅL process support system. Some were given more attention when focusing on a prototype system to be implemented in this project, the most important one being the development of a distributed and heterogeneous process engine composed of a collection of Web Services. The engine should provide adequate support for process enactment, as well as Web-based generation of to-do lists. Additionally, a simple process modelling language and tool were to be built using the already existing Pounamu meta-CASE tool. Pounamu was also to be used as the front end of IMÅL, providing user interaction with the system.

Chapter 4 - Design

4.1. Introduction

Object-Oriented Design (OOD) is concerned with the development of an object-oriented model of the system to implement the requirements identified in OOA. Hence, the focus in this chapter is on the design of the IMÅL prototype system based on the requirements outlined in Chapter 3. Firstly, software architecture for the system is presented, including architectural layers as well as general and prototype-specific architectural design diagrams and descriptions. Secondly, OOA classes are refined into OOD classes and presented in UML OOD class diagrams. UML sequence diagrams for certain aspects of the design are also included. The final section of this chapter explains some design patterns used when designing the system, and how they were applied.

4.2. Software Architecture

According to Bass et al., the software architecture of a program or computing system is defined as "the structure or structures of the system, which comprise software components, the externally visible properties of those components, and the relationships among them" [6]. Simplified, the software architecture of a system can be seen as the overall structure of the system. This section presents the software architecture of the IMÅL process tool.

In line with the functional requirements section in the previous chapter, a general conceptual architecture for the IMÅL system is proposed, followed by a presentation of the architecture specific to the prototype implementation undertaken in this project. As was the case with the prototype-specific functional requirements, the prototype-specific architecture is a narrowing of the general architecture, in order to meet the time constraints associated with this project.

4.2.1. General Architectural Design

Figure 4-1 shows the overall composition of the system in terms of three architectural layers. On top sits the user interface, provided by either Pounamu or a Web browser, and an information database and repository layer is the foundation in the system. A Web Services layer that situated in the middle serves as the connection between users and information.

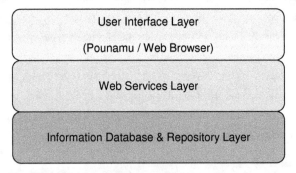

Figure 4-1. Architectural Layers in the IMÅL System.

A proposed general architecture of the IMÅL process management and support system is depicted in Figure 4-2. Each component included in the general architectural diagram is briefly explained in the following.

Process Modelling & Enactment Tool: Responsible for user interaction with the system by allowing for modelling and enactment of processes. Additionally, this component must provide a way to connect to and forward the process definition and enactment events to the main process engine service.

Main Process Engine Service: Coordinates process enactment using several services. The service uses its local <u>XML Parser</u> to interpret the process model information it receives.

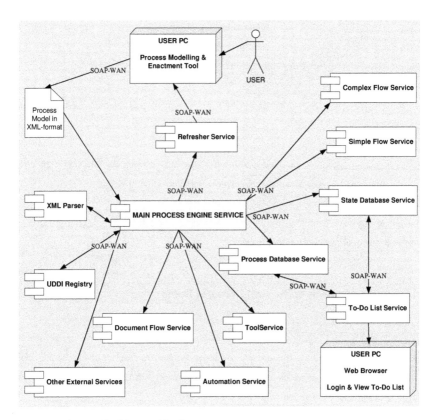

Figure 4-2. General Architectural Design.

<u>State Database Service:</u> Stores information about process state.

<u>Process Database Service:</u> Stores other process-related information.

<u>Simple Flow Service:</u> Enables simple process flow by processing enactment events.

<u>Complex Flow Service:</u> Enables complex process flow by processing enactment events.

<u>Refresher Service:</u> Stores updates to the user interface and process model resulting from the processing of enactment events.

To-Do List Service: Displays to-do lists to certified users.

Tool Service: Presents the user with third-party tools needed to perform work on stages.

Automation Service: Provides automation of tasks that need no user intervention.

Document Flow Service: Controls the flow of documents and artefacts during process enactment.

Other External Services: Includes additional external services that the main process engine might beneficially make use of, such as awareness and collaboration services.

UDDI Registry: Provides an interface for looking up and discovering external services useful to the main process engine service.

4.2.2. Prototype-Specific Architectural Design

Depicted in Figure 4-3 is the architectural design of the main components that constitute the IMÅL prototype built in this project. Based on the general descriptions in the previous section, the following provides a more in-depth explanation of each component included in the prototype along with hardware resources required.

Pounamu Process Modelling & Enactment Tool: The Pounamu software tool offers functionality to model and later enact processes based on a pre-defined process modelling tool. Providing the user interface for both modelling and enactment, Pounamu serves as the main component for user interaction with IMÅL. A process model created using the Pounamu project modeller and a pre-defined process modelling tool is stored in XML-format. The Pounamu Process Engine Component is the part of Pounamu which, when plugged in, provides a way to connect to and forward the process definition and enactment events to the main process engine service.

Main Process Engine Service: This service makes use of several external services to coordinate process enactment. It also uses its local XML Parser to interpret and create an

84

internal representation of the process model. The main process engine service and all the other services it uses together constitute the process engine in the IMÅL process tool.

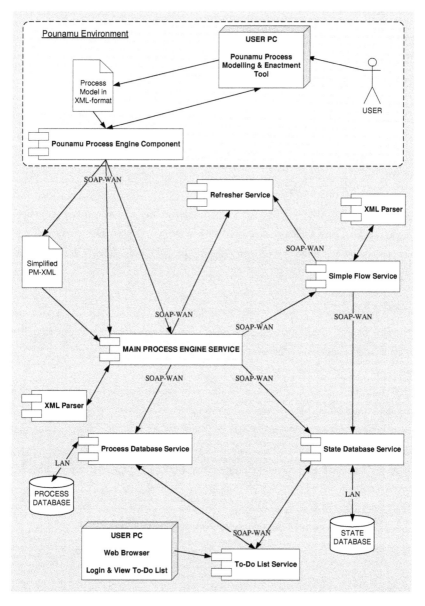

Figure 4-3. Prototype-Specific Architectural Design.

State Database Service: The state database service stores and manages information about process state. The information is stored in a database residing on the same machine as the service itself.

Process Database Service: The process database service stores other process-related information, and is built up in a similar fashion to the state database, with a database that resides locally together with the rest of the service.

Simple Flow Service: This service enables simple process flow by processing enactment events. Similarly to the main process engine service, the simple flow service has its own XML Parser that it uses to interpret and create an internal representation of the process model.

Refresher Service: The refresher service stores updates to the user interface and process model resulting from the processing of enactment events. Typical user interface updates are the adding and removing of menu items, whereas process models are updated by changing the colour of stages.

To-Do List Service: A Web browser is at the front end of the to-do list service, and provides the user interaction with the service. The service displays to-do lists as well as overall process information for users that are logged in.

Basically, the components included in the prototype-specific architecture depicted in Figure 4-3 are all crucial components in making the system work properly and provide basic process support. The components that were left out from the general architectural design shown in Figure 4-2 were considered less crucial to the working of a basic process support system, and were therefore left for future work. It is important to note, however, that the inclusion of these additional components would significantly enhance the system and its flexibility as well as increase its acceptance and possible use. Unfortunately, the time constraints associated with this project made the implementation of all components infeasible.

As explicitly shown in Figure 4-3, the components that constitute the system are Web Services spread out over different locations on a Wide Area Network (WAN). The components use Web protocols to communicate with each other, as this is standard for Web Services. Owing to the resources required being fairly low, the different services can all run on moderate sized server machines with moderate processors. However, in order to reliably support numerous processes running concurrently, high-end server machines with fast processors are preferable.

User interaction with the process support system is provided by the Pounamu Process Modelling Tool and the to-do list service respectively. Pounamu runs on a moderate-sized desktop machine with moderate processor speed, and an available connection to the Internet. The to-do list service can be accessed from any machine that supports a Web browser and has a connection to the Internet.

Figure 4-4 shows the same architectural diagram as Figure 4-3, but additionally includes important event flows between the components that make up the system. Event flows were added to the architectural design in order to illustrate how the separate components work together to support the enactment of processes.

When a process has been modelled, the Pounamu process engine component can be plugged into Pounamu at run-time (1). Thus, the process engine component becomes a part of the local environment. On start-up, the component automatically starts up the main process engine service, and sends to it a simplified version of the process model (2). The simplified version of the process model conforms to the standard format that the process engine requires. In theory, the process model can be modelled using any formalism, as long as it is translated into the appropriate format before being sent to the process engine.

The main process engine service firstly uses an XML parser to interpret the process model format and convert it into a representation of the process internal to the process engine service. It then uses the state database service to store process state information (3) and the process database service to store other relevant process information contained in the process model (4). Finally, all the other services used by the main process engine service to support process enactment are initialised. Like the main process engine service, the simple flow service uses an XML parser to create its own internal representation of the process model XML, which is forwarded to it upon initialisation (5).

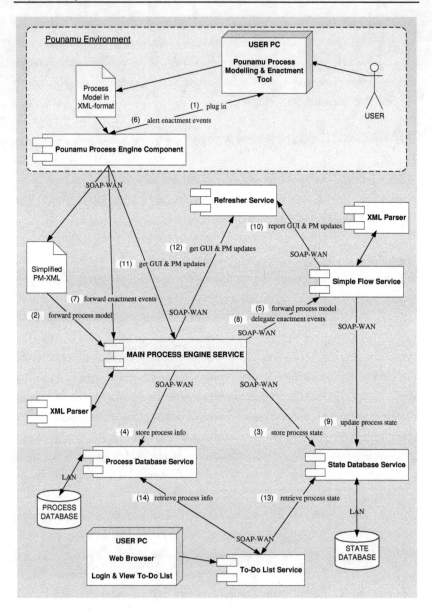

Figure 4-4. Prototype-Specific Architectural Design with Event Flows.

After initialising the environment, the process engine waits for enactment events from Pounamu. Enactment events triggered in the Pounamu project modeller are alerted to the

Pounamu process engine component (6), which forwards them to the main process engine service (7). When an enactment event is received by the main process engine, it is delegated to the simple flow service for processing (8). A possible extension to the process engine would be to add other processing components such as the complex flow service included in the general architecture in Figure 4-2. In that case, the process engine would decide which service to delegate the enactment event to, depending on the enactment event information it receives.

The simple flow service processes the enactment event by updating the state database (9) and reporting user interface and process model updates to the refresher service (10). Finally, the Pounamu process engine component retrieves necessary updates to the user interface and the process model via the process engine service (11-12), and applies them in Pounamu.

Additionally, the to-do list service can be accessed via a Web browser from any location at any time. The service validates the username and password provided by a user, and if the validation succeeds, provides the user with access to a list of individual to-do items as well as an overall view of the enacted process and its progress. To retrieve this information, the to-do list service makes use of the state database and the process database services (13-14).

4.3. OOD Class Diagrams

This section presents OOD class diagrams for the Pounamu process engine component as well as all the services that make up the process engine. The diagrams are refinements of the OOA class diagrams put forward in Section 3.6.

4.3.1. The Pounamu Process Engine Component

In the process engine component of the Pounamu tool, class ProcessEngineComp from the OOA diagram in Figure 3-8 was split up into three classes and one interface, as shown in Figure 4-5. This was done to separate the functionality into logically related classes, and to separate remote service calling from the code that works on Pounamu locally.

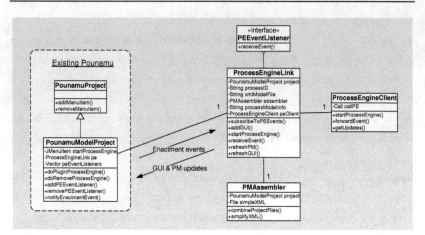

Figure 4-5. OOD Class Diagram for the Pounamu Process Engine Component.

The existing Pounamu class most important for the process engine component was, as mentioned in Section 3.6.1, the PounamuModelProject. Basically, this class serves as a programmatic representation of a user's modelling project. In the case of process support, class PounamuModelProject represents a canvas on which the user models processes and interacts with the process engine by enacting the process model. Functionality to plug in the process engine component was added to the existing PounamuModelProject class, as well as functionality to notify the component about occurring enactment events when it is plugged in. Such notification was enabled by registering the process engine component as a 'listener' to events related to process enactment.

Functionality to dynamically refresh the Pounamu user interface by adding and removing menu items, according to process enactment, was added to class PounamuProject, the more general parent class of PounamuModelProject. This functionality was added to class PounamuProject because it provides easy access to the menu bars.

The main class of the process engine component is class ProcessEngineLink, which 'links' the remote process engine service to the Pounamu software tool. It listens to process enactment events in Pounamu, by implementing the PEEventListener interface and providing a definition of the *receiveEvent* method. Class ProcessEngineLink contains a reference to the PounamuModelProject it is associated with, as well as a link to the location in which the process model definition is stored. Because Pounamu supports multiple views and multiple view-types, meaning that one process model can be spread over several view-files, the work of class PMAssembler is of considerable importance. It is responsible for assembling all the view-files into one large XML file containing the whole process model, before it simplifies this XML file into the format that the remote process engine service takes.

Class ProcessEngineLink contains methods to start up the remote process engine service, receive and forward enactment events, and update the user interface and process model according to the outcome of enactment event processing. For all interaction with the remote process engine service, class ProcessEngineLink makes use of class ProcessEngineClient. In other words, all the methods that call the remote service are encapsulated by class ProcessEngineClient.

4.3.2. The Process Engine Service

The process engine service is the main service in the process engine, and the part of the system that Pounamu's process engine component communicates with. It controls process execution by making use of several other services external to the local process engine service itself. All files and classes that make up the core process engine and reside on the same server are included in the local process engine service. The remote services can all reside on different servers.

The classes in the OOA class diagram in Figure 3-9 were refined into OOD classes in order to logically separate the functionality. At the heart of every service that make up the process engine is an interface named *ServiceNameIF*, along with an implementation class for this interface, called *ServiceNameImpl*. Methods enclosing the services' functionality are declared in the interface and implemented in the appropriate implementation class. An exception is the to-do list service, which itself is a service running on a server, and at the same time is a client of the database services. Presented in Figure 4-6 is an overall class diagram showing the composition of the process engine as a whole, including the components local to the process engine service as well as the remote services it uses.

91

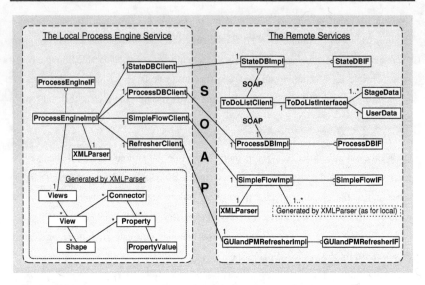

Figure 4-6. Overall Class Diagram for the Process Engine Service.

A more in-depth class diagram showing only the local components of the process engine service is depicted in Figure 4-7. This diagram shows more detail as to which functions each component offers. To further enhance comprehension, an explanation of the service is also provided. More detailed class diagrams and explanations for each one of the remote services are given in subsequent sections.

Naturally, the process engine service is the largest of the remote services that make up the process engine. Its main class is class ProcessEngineImpl, which is an implementation of the ProcessEngineIF interface. Being the main class, class ProcessEngineImpl is the class that Pounamu's process engine component connects to and communicates with. In other words, all interaction between Pounamu's process engine component and the process engine as a whole goes through this class. The class has methods for starting up and initialising the process engine service, as well as for receiving enactment events and retrieving necessary updates to the user interface and process model in Pounamu.

Class ProcessEngineImpl utilises class XMLParser to parse and generate objects from the process definition received from Pounamu, creating a representation of the process model internal to the process engine service. The objects generated by the XMLParser are instances of pre-generated classes. The generated classes include both an interface and an implementation for each class. For simplicity, only the implementation classes and their attributes and methods are shown in Figure 4-7.

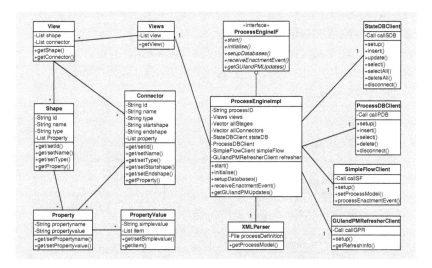

Figure 4-7. OOD Class Diagram of the Process Engine Service Locally.

Class ProcessEngineImpl makes use of class StateDBClient and class ProcessDBClient to connect to and communicate with the state database service and the process database service respectively. Similarly, class SimpleFlowClient and class GUIandPMRefresher-Client are used to connect to and communicate with the simple flow and refresher services. The database services are used by class ProcessEngineImpl to store initial state and process information when initialising the environment. Further, enactment events received from Pounamu are forwarded to the simple flow service for processing. Finally, the refresher service is used to store and retrieve user interface and process model updates.

4.3.2.1. The State and Process Database Services

The state and process database services, depicted in Figure 4-8, were designed in exactly the same fashion, with the only difference between the two being the information that is stored in the database. Process state information is stored by the state database service,

93

while the process database service stores other process-related information. The idea was that the state database is more dynamic, with frequent updates to process state when the process is being enacted, whereas the process database is more static, containing pre-defined information. This served as the initial reason for separating the database functionality into two separate services. Additionally, the design of the services provides for possible enhancement in the future. For example, supporting the editing of a process model while it is being enacted can be relatively easily achieved, as most updates will affect only the process database service, and leave the state database service to continue monitoring the enactment of processes.

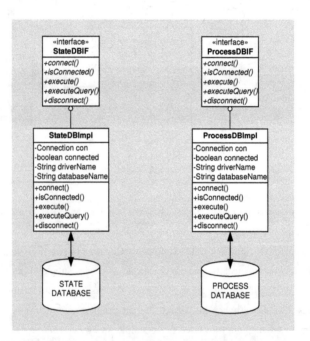

Figure 4-8. OOD Class Diagrams for the State and Process Database Services.

Consisting of only three components, namely interface StateDBIF/ProcessDBIF, class StateDBImpl/ProcessDBImpl and the actual database, the database services are relatively simple services. They contain methods to connect to and disconnect from the respective databases, as well as methods to execute insert, update, select and delete queries.

4.3.2.2. The Simple Flow Service

Interface SimpleFlowIF and its implementation class SimpleFlowImpl are at the heart of the simple flow service, shown in Figure 4-9. Initialisation of the simple flow service involves creating a local representation of the process model, and this is done in the same fashion as in the process engine service. Class SimpleFlowImpl utilises the local XMLParser to parse and generate objects from the process definition received from the process engine service. The objects generated by the XMLParser are instances of pre-generated classes identical to the ones in the process engine service. Thus, these classes were pre-generated both on the server on which the process engine service resides, and on the server on which the simple flow service resides.

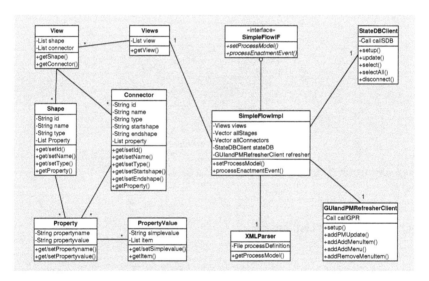

Figure 4-9. OOD Class Diagram for the Simple Flow Service.

The simple flow service needs the internal representation of the process model to be able to process enactment events, and thereby guide process flow which is the main responsibility of the service. When processing enactment events, the service makes use of class StateDBClient to connect to and update the state database according to processing results, and class GUIandPMRefresherClient to store necessary updates to the user interface and process model.

95

4.3.2.3. The Refresher Service

The refresher service, depicted in Figure 4-10, is the simplest of all the remote services. It consists of interface GUIandPMRefresherIF and its implementation class GUIandPM-RefresherImpl. The service offers functionality to store necessary updates to the user interface in terms of adding and removing menu items, as well as updates to the process model in terms of changing the colour of stages. Additionally, it offers a method to retrieve all the stored updates and reset the service.

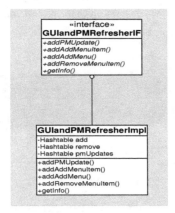

Figure 4-10. OOD Class Diagram for the Refresher Service.

4.3.2.4. The To-Do List Service

The main class in the to-do list service, shown in Figure 4-11, is the ToDoListInterface, which is responsible for processing login and view to-do list requests from the user. When processing requests, class ToDoListInterface utilises class ToDoListClient to connect to and retrieve information from the state and process database services. Another important method in class ToDoListInterface, apart from the methods to process user requests, is the *prepareTodoList* method. This method stores the information retrieved as objects of the ProcessData and UserData classes, which are responsible for managing the data related to processes and users respectively, and prepares the data that is to be displayed to the user.

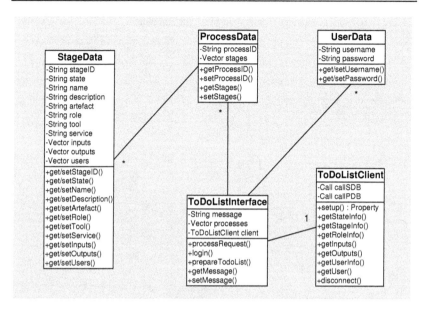

Figure 4-11. OOD Class Diagram for the To-Do List Service.

4.4. OOD Sequence Diagrams

Using sequence diagrams to capture event flows and interactions between users and the system, and between different components of the system, this section elaborates on the four use cases presented in Section 3.5. The first use case is, however, excluded here, as it is more an interaction with the existing Pounamu software tool than with the IMÅL prototype process system built in this project.

4.4.1. Plugging in Process Engine

A sequence diagram depicting the flow of events when a user decides to plug in the process engine component to a running Pounamu is shown in Figure 4-12. The scenario corresponds to Use Case 2, as outlined in Section 3.5.2.

When the user chooses to plug in the process engine from the menu bar (1), the process engine component is created (2). Upon creation, the component 'subscribes' to enactment events (3). This means that the component will be notified whenever such events occur. Additionally, process support functionality is added to the Pounamu user interface (4). The component finally starts up the remote process engine service (5), which in turn initialises the other services it uses (6), and Pounamu is ready to support process enactment.

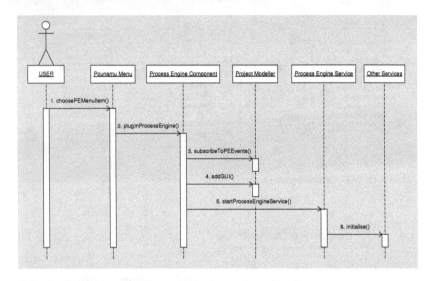

Figure 4-12. Sequence Diagram of Plugging the Process Engine Component into Pounamu.

4.4.2. Enacting a Process Model

Assuming that the user has added the process engine component to a running Pounamu, Figure 4-13 presents a sequence diagram of event flows initiated by the user enacting a process model. This situation conforms to Use Case 3, as described in Section 3.5.3.

Process enactment is initiated by the user enacting a stage in a pre-defined process model in Pounamu (1). Because the process engine component subscribed to process events when it was plugged in, it is notified about the enactment event when it occurs (2). The component forwards the event to the process engine service (3), which further delegates it to the simple flow service for processing (4). During processing, the simple flow service

98

updates the state database (5) and stores necessary updates to the process model and user interface in the refresher service (6-7). Finally, Pounamu's process engine component retrieves the updates from the refresher service via the process engine service (8-9), and applies them in Pounamu (10-11).

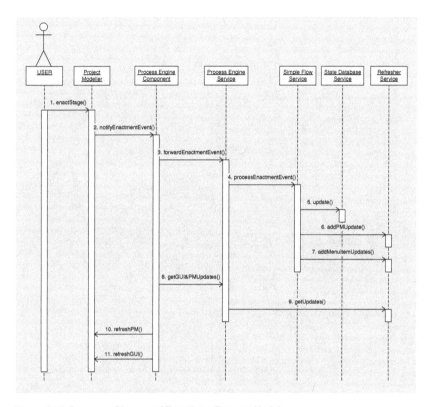

Figure 4-13. Sequence Diagram of Enacting a Process Model.

4.4.3. View To-Do List

Figure 4-14 shows a sequence diagram of important event flow in a view to-do list scenario. This scenario corresponds to Use Case 4, as outlined in Section 3.5.4.

The view to-do list scenario starts with a user pointing a Web browser to the address of the to-do list service. To be able to use the service, the user must supply his/her username and password and hit the login button (1). The login request is then processed by the to-do list service (2), by checking the username and password against a table of users in the process database (3). A message is set to indicate whether or not the login was successful (4), and the result of the login is displayed to the user (5).

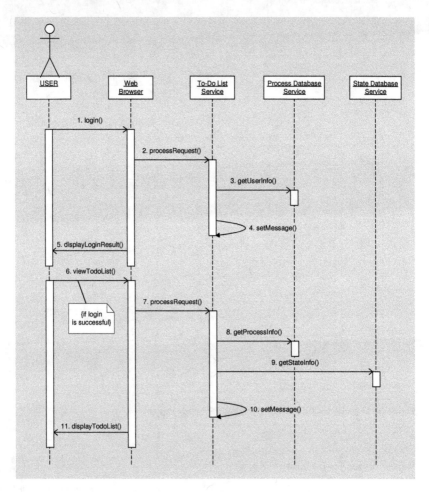

Figure 4-14. Sequence Diagram of a View To-Do List Scenario.

If the login is accepted, the user can choose a view a to-do list option (6). Again, the request is processed by the to-do list service (7). The state and process database services are used to retrieve state and process information respectively (8-9). Finally, a message is

100

set to indicate that the to-do list has been prepared (10), and the Web browser displays the list to the user (11).

4.5. Design Patterns Used

Design patterns capture the static and dynamic structures of solutions that occur repeatedly when producing applications in a particular context [25]. In other words, they refer to successful solutions to common problems in object-oriented software design [35]. Accepted categories of design patterns include creational, structural and behavioural patterns [34]. Creational patterns apply to the object creation process, structural patterns address object composition, and interaction between objects is characterised by behavioural patterns [34]. Some design patterns used when designing the process support system, and how they were applied, are outlined in the following sections.

4.5.1. The Observer Pattern

The observer design pattern is a behavioural pattern with which the intent is to "define relationship between a group of objects such that whenever one object is updated all others are notified automatically" [93].

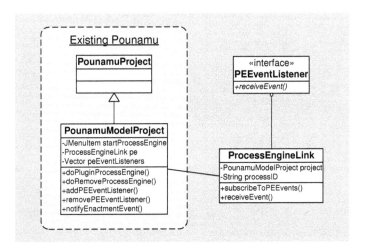

Figure 4-15. The Observer Pattern Applied in Pounamu's Process Engine Component.

101

Figure 4-15 shows how the observer pattern was applied in the design of Pounamu's process engine component. Implementing the PEEventListener interface, class ProcessEngineLink is a listener which upon initialisation subscribes to listen to enactment events. This means that it is notified of any process-related events that occur. It further provides an implementation of the *receiveEvent* method declared by PEEventListener, to determine how the component should react to enactment events. This is a flexible approach, as it allows for additional components that need to listen and react to enactment events to be easily plugged in at a later date.

4.5.2. The Decorator Pattern

The decorator pattern is another pattern used in the design of the process engine component in Pounamu. It is a structural pattern typically used in contexts where the aim is to "attach additional responsibilities or alternate processing to an object dynamically" [93]. This design pattern was therefore applied in Pounamu's process engine component in order to dynamically add functionality to Pounamu's graphical user interface.

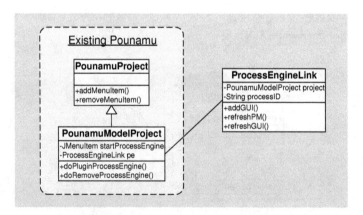

Figure 4-16. The Decorator Pattern Applied in Pounamu's Process Engine Component.

When the user plugs in the process engine component, the *addGUI* method in class ProcessEngineLink is responsible for adding extra functionality to the Pounamu menus without interfering with the execution of Pounamu itself. Similarly, the *refreshPM* and

102

refreshGUI methods update the Pounamu functionality according to the results of processed enactment events. The application of the decorator pattern in Pounamu's process engine component is shown in Figure 4-16.

4.5.3. The Façade Pattern

Façade is a structural pattern in which a complex sub-system is encapsulated using a high-level interface to simplify use and hide structural details of the sub-system [23]. This pattern was applied when designing the state and process database services in the process engine.

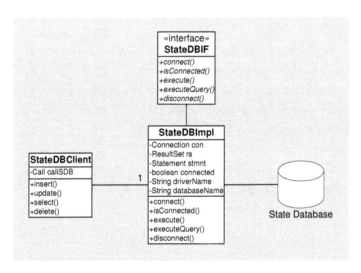

Figure 4-17. The Façade Pattern Applied in the Database Services.

The classes StateDBImpl and ProcessDBImpl encapsulate the interaction with other complex classes, such as class ResultSet, Connection and Statement. The clients that access the databases, class StateDBClient and ProcessDBClient, can therefore talk to the StateDBImpl and ProcessDBImpl classes only, and need no knowledge of their use of underlying database-related classes. Figure 4-17 shows how the façade pattern was applied in the design of the database services to appropriately shield the clients from complex sub-system components. Additionally, the pattern allows for changes in the implementation of the underlying sub-systems to be made without changing any of the client code [23].

4.5.4. The Proxy Pattern

The proxy design pattern is a structural pattern commonly used when a complex object needs to be represented by a simpler one [23]. A remote proxy is a type of proxy that provides "a local representative for an object in a different address space" [35]. Remote proxies were used frequently in the design of the process engine in this project. All remote services an environment in the engine makes use of are represented locally in the environment by remote proxies.

An example of this phenomenon is depicted in Figure 4-18, which shows how class SimpleFlowImpl in the simple flow service uses its own remote proxy GUIandPM-RefresherClient, in order to communicate with the remote refresher service itself. The remote proxy acts as an intermediate layer, hiding the fact that the refresher service resides in a different address space than the simple flow service.

As a remote proxy, class GUIandPMRefresherClient is responsible for encoding requests and sending them to the remote service, which performs processing on the remote object dependent on the received requests.

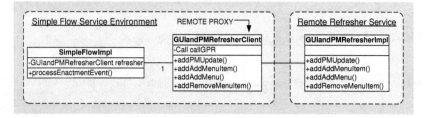

Figure 4-18. The Remote Proxy Pattern Applied in the Simple Flow Service.

An important note here is that the remote proxy design pattern is used numerous times in the design of the process support system, and not only for the simple flow service. In fact, it is applied in all parts of the system where communication with remote services is required. Hence, Figure 4-18 shows only one of many applications of the remote proxy pattern in the process support system.

4.6. Summary

The design of IMÅL comprises a decentralised software architecture incorporating a collection of Web Services. All the components most essential to supporting basic process enactment were included in the prototype implemented in this project. These components were the Pounamu process engine component, the process engine service, the process and state database services, the simple flow service, the refresher service and the to-do list service. The design of the different components was assisted by design patterns such as the observer and remote proxy patterns, and provided a solid basis for implementing the system.

Chapter 5 - Implementation

5.1. Introduction

This chapter is concerned with the conversion of the design of the IMÅL prototype presented in Chapter 4, into a working software program. It presents the underlying technologies used in the implementation, including how they work and the way in which they were applied in this project. Additionally, examples of source code and screen shots from the prototype system are presented.

5.2. Process Modelling and Enactment with Pounamu

A process modelling and enactment tool with which the user can interact was needed to serve as the front end of the IMÅL prototype. For that purpose, the in-house Pounamu modelling tool was chosen. There are several advantages of using this tool for user interaction, most of which influenced the decision.

Firstly, since the software is being developed in-house, it was easily available and free of charge. It also meant that guidance on implementation details and how to use the software was readily accessible, as the person responsible for the development of Pounamu works in the same department. Additionally, the source code was at hand, which made it reasonably simple to make changes specifically to suit this project. A final reason for using the Pounamu software tool in this project was to test it out and assist in providing feedback and suggestions for improvement of the tool.

The following elaborates on the use of the Pounamu software tool as the process modelling and enactment tool for the process support system.

5.2.1. Pounamu Process Modelling Language and Tool

Based on the visual Pounamu Process Modelling Language (PPML), the Pounamu Process Modelling Tool (PPMT) was constructed using the Pounamu tool creator. Seeing that the focus in this thesis was mainly on the process engine, the development of a simple process modelling language and tool was satisfactory. As mentioned in Section 2.6.1, a simple process modelling language should meet certain criteria. The elements described in the following were therefore used as the basic underlying elements of the PPML.

- **Stage:** A stage is a part of a process, and is described by an ID, a name, the artefact involved, the tool used, and the role responsible for performing the work associated with the stage. A stage can also be broken down into sub-stages.
- **Artefact:** An artefact is a document involved in a process stage. It is described by a name.
- **Tool:** A tool is a program or other resource used to perform the work contained in a stage. It is described by a name.
- **Role:** A role is a categorised description of someone responsible for completing the work included in a stage, such as project manager or software designer. It is described by a name.
- **Actor:** An actor is a human user that fills a role, and it is described by the user's name. As processes are automated, human actors can alternatively be substituted with computerised agents.
- **Flow:** A flow defines the flow of a process by connecting two stages. It can be described by a name.

When building the actual tool, a more specific representation was required in order to provide an appropriate syntax. This enables the definition of for example different types of stages, such as start, stop and work stages. The specific process modelling elements included in PPML, which consist of Pounamu shapes and connectors, are presented in the following. Each element is described in terms of its name and representation, its description, its attributes, and its initial visual appearance.

- **StartStage – Shape:**
  - o Description: The first stage in a process (contains no work).
  - o Attribute: name = 'Start' (fixed)
  - o Appearance:

- **StopStage – Shape:**
  - o Description: The last stage in a process (contains no work).
  - o Attribute: name = 'Finish' (fixed)
  - o Appearance:

- **BaseStage – Shape:**
  - o Description: A basic stage in a process (contains work to be completed).
  - o Attributes:
    - ID = the stage identifier
    - name = the name of the stage
    - description = a description of the work involved
    - role = the role responsible for performing the work
    - artefact = the artefact involved
    - tool = the tool to be used
    - service = the service instance responsible for processing this stage
    - inputs = a list of possible inputs to the stage
    - outputs = a list of possible output results from the stage
  - o Appearance:

- **LinkStage – Shape:**
  - o Description: Joins two or more basic stages which all flow into one other basic stage* (contains no work).
  - o Attribute: operator = AND/OR/XOR
  - o Appearance: ○
    * Note: An OR operation can also be modelled by two or more connectors going into the same stage, without using the LinkStage.

109

- **ChoiceStage – Shape:**
    - o Description: Splits the process flow from one basic stage into flows to two or more basic stages depending on a condition* (contains no work).
    - o Attribute: value = the value of the condition
    - o Appearance: ⟨⟩

        * Note: A flow going from one stage to two or more other stages can also be modelled by multiple connectors from one stage to various other stages, without using the ChoiceStage.

- **RoleActor – Shape:**
    - o Description: An association between a role and actors (contains no work).
    - o Attributes:
        - ▪ role = the name of the role
        - ▪ actors = a list of the actors filling this role
    - o Appearance: ☐

- **Flow – Connector:**
    - o Description: A representation for process flow.
    - o Attribute: name = the name of the flow (optional)
    - o Appearance: ⟶

Figure 5-1 shows a screen shot of the PPMT. In the panel on the left hand side, a tree-manager contains all the elements contained in the process modelling tool. These include six shapes, one connector, one meta-model view, and one view-type. The main modelling area, shows the only view-type defined in the tool, which includes all the elements mentioned. A process modelled using PPMT is depicted in Figure 5-2. As described in Section 3.5.1, the scenario captured in this process model is the one in which new functionality is to be added to an existing software system. Any process can in reality be modelled in multiple views. However, the screen shot included here shows the whole process model, including role-actor associations, in one view in order to give a clear picture of the process scenario that is being modelled. The tree in the left-hand panel contains all the elements that make up the process model, while the main panel shows the visual representation of the process.

110

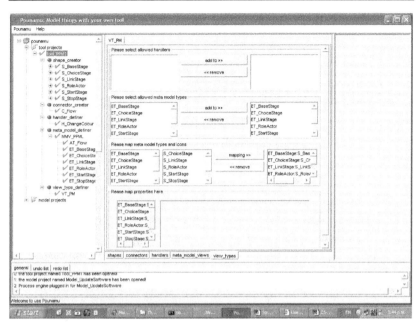

Figure 5-1. The Pounamu Process Modelling Tool.

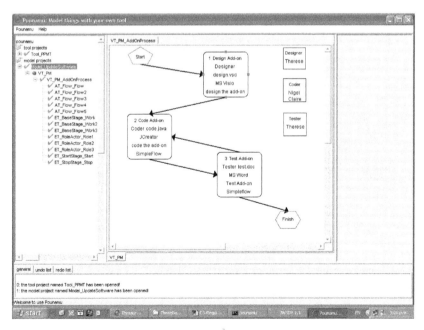

Figure 5-2. A Process Modelled using the Pounamu Process Modelling Tool.

5.2.2. XML as the Process Model Format

In the Pounamu environment, XML technology is used as the saving and loading format. XML stands for eXtensible Markup Language and is a simple and flexible tag-based text format derived from another markup language, namely SGML (Standard Generalized Markup Language [96]) [102]. XML was designed for ease of implementation and for interoperability with both SGML and HTML (HyperText Markup Language [99]), and plays an important role in the exchange of data on the Internet and elsewhere [102].

Important advantages of XML include both platform and system independence, enabling it to be used on any computer [97]. Further, it is flexible and customisable, permitting users to create their own tags, and any XML-aware software are able to work with customised tags. When working with XML documents, style sheets defining how to display the markup are separate, allowing it to be displayed in numerous customised ways [97]. In other words, XML separates data from presentation. Additionally, a wide range of software tools to work with XML are available. A disadvantage of XML, on the other hand, is that fairly recent software is required to be able to read and understand it.

```
<!-- ProcessModel.dtd: defines the valid structure of a process model xml file  -->
<!-- root element: processmodel  -->
<!ELEMENT processmodel (view)*>
<!ELEMENT view (shape|connector)*>
<!ELEMENT shape (name,type,id,property*)>
<!ELEMENT connector (name,type,id,startshape,starthandler,endshape,endhandler,property*)>
<!ELEMENT property (propertyname,propertyvalue)>
<!ELEMENT propertyvalue (simplevalue|item*)>
<!ELEMENT name (#PCDATA)>
<!ELEMENT type (#PCDATA)>
<!ELEMENT id (#PCDATA)>
<!ELEMENT startshape (#PCDATA)>
<!ELEMENT starthandler (#PCDATA)>
<!ELEMENT endshape (#PCDATA)>
<!ELEMENT endhandler (#PCDATA)*>
<!ELEMENT propertyname (#PCDATA)*>
<!ELEMENT simplevalue (#PCDATA)*>
<!ELEMENT item (#PCDATA)*>
```

Figure 5-3. DTD for Process Model Created Using the Pounamu Process Modelling Tool.

XML is used in order to save to disk everything from tool and model projects to simple shapes and connectors created using Pounamu. Thus, when a process model is saved in Pounamu, an XML file describing the process model is saved to disk. Because the Pounamu software tool already offered this functionality, it made sense to utilise it when building the IMÅL prototype. Additionally, the XML format is beneficial when dealing with distributed systems, as it can describe relatively complex types using simple text. The process support system built in this project is of a highly distributed character. Seeing that sufficient software to work with XML was available in this project, its various advantages made it suitable as the format in which to represent process models.

The permitted structure of an XML document can be defined in a grammar called Document Type Definition (DTD) [102]. Figure 5-3 displays the DTD for a process model created using PPMT.

5.2.3. Enabling Process Enactment in Pounamu

A process engine component was added to the existing Pounamu tool in order to provide a way for the user to enact processes and activate the process engine via the Pounamu user interface. As Pounamu is implemented in the Java programming language, using Java also for its process engine component was natural.

To enable the adding of the process engine component to Pounamu in a component-based fashion [41], a PEEventListener interface was added to the original Pounamu code. The interface contains a *receiveEvent* method, which is called when a process-related event occurs in Pounamu. Any component wanting to be notified about process-related events, implements this interface, and specifies in its *receiveEvent* method the code that should be executed when such events occur. Figure 5-4 displays the code for the PEEventListener interface.

```
/** PEEventListener.java: interface for listening to process-related events */
public interface PEEventListener {
  public void receiveEvent(ActionEvent event, DefaultMutableTreeNode node);
}
```

Figure 5-4. Interface PEEventListener.

When plugged in, Pounamu's process engine component wants to listen and react to process-related events. Thus, the main class in the process engine component, class

113

ProcessEngineLink, implements the PEEventListener interface, and contains an implementation of the *receiveEvent* method.

A screen shot of how the plugging in of Pounamu's process engine component is done is shown in Figure 5-5. When the user plugs in the process engine component, class ProcessEngineLink is instantiated, and it immediately subscribes to listen to process events. Further, it adds menu items for process enactment support to Pounamu. It also uses class PMAssembler to assemble and simplify the process model into a simpler XML format than the one stored by Pounamu. Class PMAssembler uses a pre-defined XSL (eXtensible Stylesheet Language [103]) style sheet to perform the simplification of the process model XML, extracting only the information important for the process engine. The final task of class ProcessEngineLink upon initialisation is to start up the remote process engine service, and send to it the simplified process model. This is done via class ProcessEngineClient. After completing all these steps, Pounamu is ready to handle enactment of the process model.

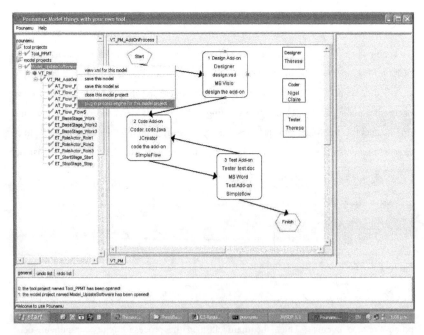

Figure 5-5. Plugging the Process Engine Component into Pounamu.

Every time a process-related event occurs in the running Pounamu, the *receiveEvent* method receives the event and forwards it to the process engine service. The enactment event is forwarded in a simple type supported by the Web Services technology, in this case a String. Further, the method retrieves the updates to the process model and user interface, and applies them in Pounamu via the *refreshPM* and *refreshGUI* methods.

This scenario is depicted in the screen shots in Figure 5-6, which show a user finishing work on a stage (Screen Shot 1) followed by the colours of the process model being updated according to the enactment (Screen Shot 2). Red means that a stage is ready to be enacted, blue means that it is currently being worked on, and a stage that is finished turns green. A whole process is completed when all the stages, including the start and stop stages, are green.

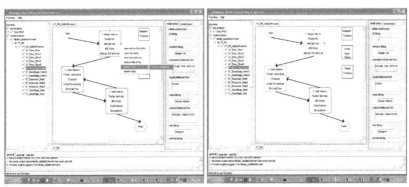

Figure 5-6. The Enactment of a Process and the Result of Processing it.

As mentioned, class ProcessEngineLink is the main class in the component of Pounamu that acts as a link to the remote process engine system. The class communicates only with the process engine service, which in turn is responsible for all communication with the external services it uses. Implementing the system in such a fashion contributed to completely separate the services the process engine utilises from the Pounamu software tool itself. Because of this separation, the user only perceives one external component, namely the process engine service, and need know nothing about the other underlying components. Figure 5-7 shows the most essential methods of class ProcessEngineLink, including the *receiveEvent* method.

115

```
/** ProcessEngineLink.java: responsible for linking Pounamu to the process engine */
public class ProcessEngineLink implements PEEventListener {
  PounamuModelProject modelProject;
  String processModelFile;
  ProcessEngineClient peClient;
  String processID;
  boolean refresherOn = true;

  /** constructor */
  public ProcessEngineLink(PounamuModelProject modelProject, String processModelFile) {
    this.modelProject = modelProject;
    this.processModelFile = processModelFile;
    subscribeToPEEvents();
    String pmInfo = assembleModelProject();
    peClient = new ProcessEngineClient();
    processID = peClient.startProcessEngine(processName, pmInfo);
  }

  /** subscribes to listen to process-related events in Pounamu */
  public void subscribeToPEEvents() {
    if(modelProject != null) {
      modelProject.addPEEventListener(this);
    }
  }

  /** assemble and simplify the files that make up the process definition */
  public String assembleModelProject() {
    PMAssembler modelAssembler = new PMAssembler(modelProject, processModelFile);
    modelAssembler.combineProjectFiles();
    File processModel = modelAssembler.simplifyXML();
    StringBuffer pmInfo = new StringBuffer();
    try{
      BufferedReader in = new BufferedReader(new FileReader(processModel));
      while(in.ready()) {
        pmInfo.append(in.readLine());
        pmInfo.append("\n");
      }
    }
    catch(Exception e) {
      System.out.println("Exception when assembling model project: " + e.toString());
    }
    return pmInfo.toString();
  }

  /** forward received event to engine and update GUI and PM according to result */
  public void receiveEvent(ActionEvent event, DefaultMutableTreeNode node) {
    if(event != null && node != null) {
      String stageName = node.toString();
      String enactmentEvent = event.getActionCommand();
      peClient.forwardEvent(stageName, enactmentEvent);
    }
    if(refresherOn) {
      Hashtable h = peClient.getRefreshInfo();
      Hashtable remove = (Hashtable)h.get("removeGUI");
      Hashtable add = (Hashtable)h.get("addGUI");
      Hashtable pmUpdates = (Hashtable)h.get("pmUpdates");
      refreshPM(pmUpdates);
      refreshGUI(remove, add);
    }
  }
}
```

Figure 5-7. The Most Essential Methods of Class ProcessEngineLink.

The system further allows for multiple processes to be run concurrently by plugging in a new process engine for each process model. Each process is assigned a unique identifier by the main process engine service, and this identifier is included in the databases that store information about processes and their states. Hence, the process engine service can distinguish between processes and make updates to the correct ones when enactment events are processed.

5.3. The Process Engine Built as an Orchestra of Web Services

In order to distribute the process engine components and allow easy dynamic integration of environment components, the service-oriented architecture depicted in Figure 4-3 was employed. Web Services technology, which is a reasonably new and emerging technology, was chosen as the technology with which to implement the service-oriented process engine. The World Wide Web Consortium (W3C) describes a Web Service as "a software system designed to support interoperable machine-to-machine interaction over a network" [101]. Other systems typically interact with a Web Service via HTTP (Hypertext Transfer Protocol [98]) using XML-based messages [101].

The implementation of the service-oriented architecture in Figure 4-3 sees several components implemented as separate Web Services with their own functionality, creating a process engine from an orchestra of Web Services. Each service can run on a separate machine, producing a highly distributed environment. However, even though the components are separate distributed Web Services, they are all necessary in order for the process engine as a whole to work properly.

The Web Services based architecture employed combines the best features of service-oriented architecture with the Internet, and all communication between services utilises XML. However, XML alone is not sufficient for communication. Figure 5-8 shows three Web Services technologies that provide standard formats and protocols for XML interpretation, and how they work together to enable communication between consumers and Web Services. A standard way of describing Web Services is defined by the Web Services Description Language (WSDL), whereas the Universal Description, Discovery and Integration (UDDI) technology provides ways to register and discover them [89].

Further, the Simple Object Access Protocol (SOAP) defines a standard communication protocol for Web Services [89].

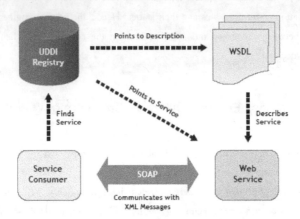

Figure 5-8. Web Services Technologies (from [89]).

The consumer of a Web Service can be a human user, an application program, or another Web Service. In the process engine, the service consumer is always an application or another Web Service. In other words, there is no direct communication between the user of the process support tool and any of the Web Services that make up the process engine.

Web Services offer many benefits over other Internet-based protocols. Being both language and platform independent, they allow for implementation in a wide range of programming languages and deployment on numerous different platforms [89]. As a result, the usual constraints of other protocols such as for example CORBA [65] and RMI [83] are eliminated. Web Services support loosely coupling of components in a system, and are therefore more amenable to necessary alterations caused by changes in the environment in which they are used. Further, their component-based nature makes Web Services easy to integrate with other systems and easy to reuse [89].

Because of the advantages they offer over similar technologies, Web Services were chosen to facilitate component distribution in this project. By using Web Services technology, a basis is also provided for additional components to be easily integrated into the system in the future, to enhance its overall functionality and usability.

The Java Web Services Developer Pack (JWSDP) was used for the development of the Web Services that constitute the process engine. JWSDP includes a version of Tomcat [4], which was used as the servlet container for the Web Services to run in.

5.3.1. Java as the Programming Language

As the platform on which to build the process engine part of the process support system, the Java platform [79] was chosen. This involved using the Java programming language for coding purposes along with already existing tools and methods provided by the Java platform. Java has several advantages that influenced the decision to use it in the implementation of the process engine.

Firstly, as well as being reasonably intuitive, Java is a highly object-oriented language [82]. As the requirements analysis and design of the process support system were carried out in an object-oriented manner, an object-oriented programming language for implementation was favourable. Other important advantages of Java include its portability, enabling it to run on any platform, as well as its robustness and good security [82]. Java is also multi-threaded, allowing applications to have multiple threads with different responsibilities running at the same time [82]. This served as a considerable advantage when building a process engine that might need to deal with and respond to several different events simultaneously.

Another feature of Java, which was highly essential to the process engine, is its extensive support for networking and distributed systems. As component distribution is a crucial issue in the design of the process engine, choosing Java as the implementation language was natural.

Java also offers comprehensive functionality for working with XML and developing Web Services. According to Sun Microsystems, XML and Java technology are perceived as "ideal building blocks for developing Web Services and applications that access Web Services" [81]. Java's compatibility with XML and Web Services was very beneficial when implementing the process engine, as its underlying architecture is both XML-based and Web Services oriented.

119

There are also some disadvantages associated with using Java as the development base. For example, because it employs the Java Virtual Machine to interpret Java binary code [79], it is notably slower than similar technologies that offer natively compiled code directly. However, the advantages associated with using Java to work with XML and Web Services in a distributed object-oriented environment decreased the importance of the speed issue in the context of this project. Additionally, seeing that the process modelling and enactment part of the IMÅL process system (Pounamu) is built using Java, it made sense to utilise Java also for the process engine, in order to conveniently integrate the two.

5.3.2. JAXB to Generate Classes and Parse Process Model XML

Java Architecture for XML Binding (JAXB) provides tools and methods for automatic mapping between XML documents and Java objects [81]. It can be used to compile an XML Schema into Java classes, which together with the binding framework enable operations such as un-marshalling and marshalling to be performed on an XML document [81]. The compilation is done by a tool called the XJC binding compiler. Figure 5-9 shows how JAXB works in the context of an application.

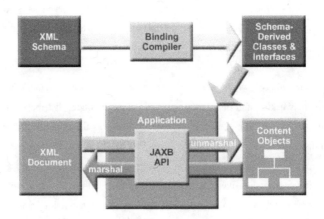

Figure 5-9. Java Architecture for XML Binding (from [81]).

In the process engine environment developed in this project, JAXB was used in order to generate Java classes from an XML Schema [100]. The XML Schema that the class

```
<!-- ProcessModel.xsd: describes the format of process models for code generation -->
<xsd:schema xmlns:xsd="http://www.w3.org/2001/XMLSchema">
<!-- root element: processmodel -->
<xsd:element name="processmodel" type="Views"/>

<!-- element: Views -->
<xsd:complexType name="Views">
  <xsd:sequence>
      <xsd:element name="view" minOccurs="1" maxOccurs="unbounded" type="View"/>
  </xsd:sequence>
</xsd:complexType>

<!-- element: View -->
<xsd:complexType name="View">
  <xsd:sequence>
      <xsd:element name="shape" minOccurs="1" maxOccurs="unbounded" type="Shape"/>
      <xsd:element name="connector" minOccurs="1" maxOccurs="unbounded" type="Connector"/>
  </xsd:sequence>
</xsd:complexType>

<!-- element: Shape -->
<xsd:complexType name="Shape">
  <xsd:sequence>
    <xsd:element name="name" type="xsd:string"/>
    <xsd:element name="type" type="xsd:string"/>
    <xsd:element name="id" type="xsd:string"/>
    <xsd:element name="property" minOccurs="0" maxOccurs="unbounded" type="Property"/>
  </xsd:sequence>
</xsd:complexType>

<!-- element: Connector -->
<xsd:complexType name="Connector">
  <xsd:sequence>
    <xsd:element name="name" type="xsd:string"/>
    <xsd:element name="type" type="xsd:string"/>
    <xsd:element name="id" type="xsd:string"/>
    <xsd:element name="startshape" type="xsd:string"/>
    <xsd:element name="starthandler" type="xsd:string"/>
    <xsd:element name="endshape" type="xsd:string"/>
    <xsd:element name="endhandler" type="xsd:string"/>
    <xsd:element name="property" minOccurs="0" maxOccurs="unbounded" type="Property"/>
  </xsd:sequence>
</xsd:complexType>

<!-- element: Property -->
<xsd:complexType name="Property">
  <xsd:sequence>
    <xsd:element name="propertyname" type="xsd:string"/>
    <xsd:element name="propertyvalue" type="PropertyValue"/>
  </xsd:sequence>
</xsd:complexType>

<!-- element: PropertyValue -->
<xsd:complexType name="PropertyValue">
  <xsd:sequence>
    <xsd:choice>
      <xsd:element name="simplevalue" type="xsd:string"/>
      <xsd:element name="item" minOccurs="0" maxOccurs="unbounded" type="xsd:string"/>
    </xsd:choice>
  </xsd:sequence>
</xsd:complexType>
</xsd:schema>
```

Figure 5-10. The XML Schema Used by the XJC Compiler.

121

generation was based on is depicted in Figure 5-10. It describes the components contained in a process model XML file. It conforms closely to the DTD for process models presented in Figure 5-3.

When generated, the classes were used as a part of the application itself. When IMÅL is running, both the process engine service and the simple flow service use JAXB to un-marshal the process model XML they receive, using the generated classes to create a representation of the process model internal to the respective services. Details as to how the un-marshalling is done are more closely investigated in the next section, which looks at the process engine service specifically.

There were several reasons behind choosing JAXB to deal with XML in the process engine environment. Firstly, the technology perfectly supports the functionality needed in the process engine, namely the generation of classes from a specification (XML Schema) and the conversion of XML documents into Java representations. This simplification of access to the XML from the Java program is extremely beneficial, as it provides the Java program with a direct source to the XML data [81]. JAXB is also more memory-efficient than other similar technologies, and can be customised in a variety of ways [81]. Finally, JAXB is a flexible approach, allowing a series of different un-marshal, marshal and validate operations [81].

5.3.3. The Process Engine Web Service

Several services constitute the process engine in the process support system. Using other external services to coordinate process enactment, the process engine service is the main service in the process engine. The other external services are the state database service, the process database service, the simple flow service, the refresher service and the to-do list service.

Although the services contain different functionality, the core of all the Web Services that together make up the process engine is implemented in a similar fashion. To show an example of how the services are implemented, this section closely inspects the process

122

engine service. As it is the main service in the system, additional specific information about the process engine service is also included here.

```
/** ProcessEngineIF.java: interface for the main process engine class */
public interface ProcessEngineIF extends Remote {
  public Hashtable getGUIandPMUpdates() throws RemoteException;
  public void receiveEnactmentEvent(String stageName, String enactmentEvent) throws RemoteException;
  public void start(String pmInfo) throws RemoteException;
  public void initialise() throws RemoteException;
  public void setupDatabases() throws RemoteException;
}
```

Figure 5-11. Interface ProcessEngineIF.

```
/** ProcessEngineImpl.java: responsible for coordinating all process enactment */
public class ProcessEngineImpl implements ProcessEngineIF {
  Views views;
  StateDBClient stateDB;
  ProcessDBClient processDB;
  GUIandPMRefresherClient refresher;
  SimpleFlowClient simpleFlow;
  String processID;
  Vector allStages;
  Vector allConnectors;

  /** constructor */
  public ProcessEngineImpl() {}

  /** receive enactment event and forward it to a processing instance */
  public void receiveEnactmentEvent(String stageName, String enactmentEvent) {
    if(stageName.startsWith("ET_StartStage")) {
      Hashtable enactmentResult = simpleFlow.processEnactmentEvent(stageName, enactmentEvent);
    }
    else {
      for(int i = 0; i < allStages.size(); i++) {
        Shape stage = (Shape)allStages.elementAt(i);
        if(stage.getName().equals(stageName)) {
          List properties = stage.getProperty();
          for(Iterator j = properties.iterator(); j.hasNext();) {
            Property property = (Property)j.next();
            if(property.getPropertyname().equals("service")) {
              String service = property.getPropertyvalue().getSimplevalue();
              if(service.toLowerCase().equals("simpleflow")) {
                Hashtable enactmentResult = simpleFlow.processEnactmentEvent(stageName,
enactmentEvent);
              }
            }
          }
        }
      }
    }
  }
}
```

Figure 5-12. The *receiveEnactmentEvent* Method of Class ProcessEngineImpl.

To conform to the requirements of Web Services, the core of the process engine service consists of an interface, ProcessEngineIF, and a class ProcessEngineImpl that implements

123

this interface. Interface ProcessEngineIF extends Remote and lists all the remote methods the service offers. As required by Web Services technology, each method throws a RemoteException if it fails. Figure 5-11 shows the code for interface ProcessEngineIF.

Class ProcessEngineImpl implements the ProcessEngineIF interface, and provides the implementation of all the methods in the interface. The implementation of the *receiveEnactmentEvent* method, which is essential to coordinating process enactment, is shown in Figure 5-12.

```
/** XMLParser.java: responsible for parsing xml file containing process definition */
public class XMLParser {
  File processDefinition = new File("processdefinition.xml");

  /**  constructor */
  public XMLParser(String pmInfo) {
    try{
      BufferedWriter out = new BufferedWriter(new FileWriter(processDefinition));
      out.write(pmInfo);
      out.flush();
      out.close();
    }
    catch(Exception e) {
      System.out.println("Exception in XMLParser constructor: " + e.toString());
    }
  }

  /**  parse process definition file */
  public Views getProcessModel() {
    try {
      JAXBContext jc = JAXBContext.newInstance("ProcessEngine.base.generated");
      Unmarshaller u = jc.createUnmarshaller();
      Views views = (Views)u.unmarshal(new FileInputStream(processDefinition));
      return views;
    }
    catch(Exception e) {
      System.out.println("Exception in XMLParser constructor: " + e.toString());
    }
    return null;
  }
}
```

Figure 5-13. Class XMLParser.

Depicted in Figure 5-13, class XMLParser is another important class in the process engine service. It is used by class ProcessEngineImpl to un-marshal the process model information the service receives, in order to create an internal representation of the process model. The un-marshalling is based on the JAXB-generated classes described in the previous section, and creates instances of these classes according to the process model definition. In other words, all the elements contained in the process model XML are converted into Java objects and attributes. These objects and attributes make up the internal representation of the process model, and are used as a part of the process engine service application.

```java
/** GUIandPMRefresherClient.java; responsible for communication with refresher service */
public class GUIandPMRefresherClient {
  private Call callGPR;
  private String endpointGPR = "http://localhost:8080/service5/refresher";
  private String qnameServiceGPR = "Refresher";
  private String qnamePortGPR = "RefresherIF";
  private String BODY_NAMESPACE_VALUE_GPR = "http://com.test/wsdl/Refresher";
  private String ENCODING_STYLE_PROPERTY = "javax.xml.rpc.encodingstyle.namespace.uri";
  private String NS_2 = "http://java.sun.com/jax-rpc-ri/internal";
  private String URI_ENCODING = "http://schemas.xmlsoap.org/soap/encoding/";
  private QName hashtable = new QName(NS_2, "hashtable");

  /** constructor */
  public GUIandPMRefresherClient() {
    setup();
  }

  /** set up client to use service */
  public void setup() {
    try {
      ServiceFactory factoryGPR = ServiceFactory.newInstance();
      Service serviceGPR = factoryGPR.createService(new QName(qnameServiceGPR));
      QName portGPR = new QName(qnamePortGPR);
      callGPR = serviceGPR.createCall(portGPR);
      callGPR.setTargetEndpointAddress(endpointGPR);
      callGPR.setProperty(Call.SOAPACTION_USE_PROPERTY, new Boolean(true));
      callGPR.setProperty(Call.SOAPACTION_URI_PROPERTY,"");
      callGPR.setProperty(ENCODING_STYLE_PROPERTY, URI_ENCODING);
    }
    catch (Exception ex) {
      System.out.println("Exception in refresher setup: " + ex.printStackTrace());
    }
  }

  /** get refresh info */
  public Hashtable getRefreshInfo() {
    Hashtable info = null;
    try {
      callGPR.removeAllParameters();
      callGPR.setReturnType(hashtable);
      callGPR.setOperationName(new QName(BODY_NAMESPACE_VALUE_GPR, "getRefreshInfo"));
      info = (Hashtable)callGPR.invoke(null);
    }
    catch (Exception ex) {
      System.out.println("Exception in get refresh info: " + ex.printStackTrace());
    }
    return info;
  }
}
```

Figure 5-14. Class GUIandPMRefresherClient.

The remaining segments of the process engine service are the different clients that communicate with the external Web Services. StateDBClient, ProcessDBClient, Simple-FlowClient and GUIandPMRefresherClient are all classes that class ProcessEngineImpl

125

delegates the dealing with the respective services to. Separating the different clients from each other and from the core of the process engine service makes the composition of the service well arranged and easy to understand, and encapsulates all the remote service calling away from the core functionality offered by the service itself. Figure 5-14 shows an example of one of the clients in the process engine service, namely class GUIandPMRefresherClient.

5.3.4. Microsoft Access for the Database Services

Research has shown that representing dynamic data using XML format is undesirable, as updates can be cumbersome and slow [104]. On that basis, Microsoft Access [58] was selected as the underlying database in which to store process state and other process-related information. A Java implementation was easily adapted to meet the needs of this project.

Although Access is less scalable and stable than other similar systems, like SQL Server [60] and Oracle [67], it is sufficient for use in the prototype developed in this project. This is because the amount of data that needs to be stored is relatively small, requiring only simple databases with a few tables and fields. Additionally, the databases in the system reside on the same server as the application that accesses them, avoiding potential performance problems associated with remote computing. However, for future support of multiple processes and users accessing the system concurrently, and to provide better scalability and stability, the use of better and more robust database technologies should be implemented. In the meantime, Microsoft Access is less resource-consuming, and a cost-efficient and applicable solution.

The state database, depicted in Figure 5-15, contains only one table, namely the StageState table. ProcessID, stageID and state are the three fields in the StageState table. The state field can have the values not ready, ready, enacted, paused or completed.

The process database, shown in Figure 5-16, has six tables named process, stage, user, input, output and role. Information about processes, stages and users is stored in the process, stage and user tables respectively. The input and output tables simply store

126

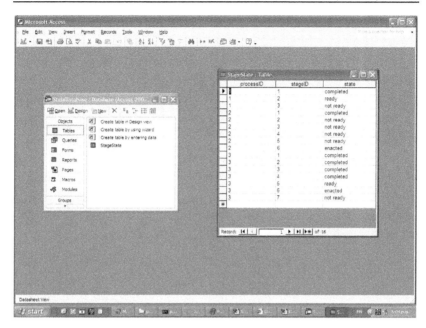

Figure 5-15. The State Database.

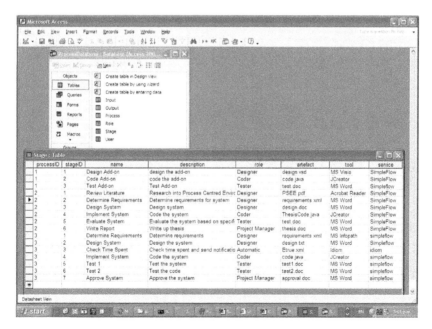

Figure 5-16. The Process Database.

127

possible input and output states associated with process stages, while the role table stores role-actor associations in processes. The stage table, which is opened in the screen shot in Figure 5-16, is the largest table in the process database, and its fields are processID, stageID, name, description, role, artefact, tool and service.

5.3.5. JSP-Based To-Do List Service

In the IMÅL prototype system, Java Server Pages (JSP) technology was utilised to provide users with access to to-do lists for enacted processes. JSP technology is an extension of Java Servlets [84]. It is a platform independent, thin-client technology that allows for rapid development and easy maintenance of dynamic Web pages [57].

A JSP page is a Web page containing dynamic Java processing code embedded in static HTML code. The Java code in a JSP page typically accesses data from server-side components such as JavaBeans. When the page is displayed in a Web browser, both the static presentation code and the dynamic content are displayed. Figure 5-17 shows how data access is achieved from a JSP page.

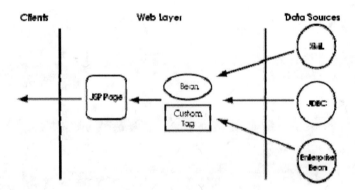

Figure 5-17. Accessing Data from a JSP Page (from [80]).

There are several advantages of using JSPs over Java Servlets, as they have all the benefits of Java Servlets and more. JSPs can utilise the full power of the Java language and libraries, and is therefore very powerful. They are also reasonably efficient, seeing that they are compiled into byte code and executed without the need for interpretation.

128

Presentation code is separated from application processing logic, enabling designers to easily change page layouts without altering the dynamic content of underlying applications [57]. The separation also contributes to hide complex application logic and enhance code reusability. Overall, JSP technology provides a convenient way to create Web applications that connect to server-side Java components [57].

Seeing that it was used as the servlet container for the other Web Services in the process support system, Tomcat was used also to run the to-do list JSP service. Figure 5-19 shows some of the code for ToDoList.jsp, which is the JSP page that displays to-do lists to users that are logged in to the service.

An example of a to-do list displayed to a certified user by the to-do list service is shown in Figure 5-18. It first lists the stages the user is responsible for, along with essential information about these stages. Further, information about the whole process is displayed with the intent of giving the user an idea of the overall state of the process.

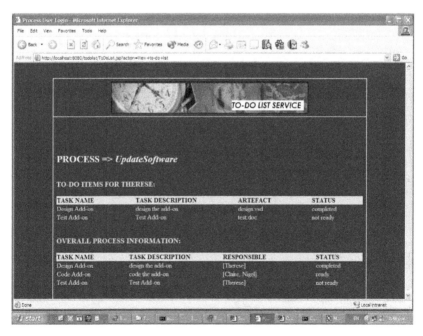

Figure 5-18. Example of a To-Do List Displayed to certified Users.

129

```
<!-- ToDoList.jsp: processes view to-do list requests -->
<HTML>
  <HEAD><TITLE>Process View To-Do List</TITLE>
    <!-- refresh page every 20 seconds -->
    <meta http-equiv="refresh" content="20"></HEAD>
  <body bgcolor="#532821" text="#FFCC99" topmargin="0" leftmargin="0">
    <!-- declare java beans -->
    <jsp:useBean id='todoListInterface' scope='session' class='ToDoListInterface'/>
    <jsp:useBean id='userdata' scope='session' class='UserData'/>
    <jsp:setProperty name="userdata" property="*" />
    <%@ page import="ToDoListInterface"
    @ page import="UserData"
    @ page import="StageData"
    @ page import="java.util.*"
    <!-- process request -->
    todoListInterface.processRequest(request.getParameter("action"), userdata);
    String message = todoListInterface.getMessage();
    if(message.equals("View to-do list")) { %>
    <div align="center">
      <!-- create tables -->
      <table border="1" width="800" height="300" bordercolor="#FFCC66" cellpadding="0"
cellspacing="0">
        <tr>
          <td width="54%" height="78" align="center"><img border="0" src="title.gif" align="center"
width="500" height="78"></td>
        </tr>
        <tr>
          <td width="54%" height="230" align="center">
            <table width="780" cellspacing="0"><tr><font color="#000000"><td bgcolor="#FFCCCC"
width="140"><b>TASK NAME</b></td><td bgcolor="#FFCCCC" width="180"><b>TASK DESCRIPTION</b></td><td
bgcolor="#FFCCCC" width="130"><b>ARTEFACT</b></td><td bgcolor="#FFCCCC"
width="90"><b>STATUS</b></td></font></tr>
              <!-- get process info -->
              <%Vector processes = todoListInterface.getProcesses();
              for(int m = processes.size()-1; m > -1; m--) {
                ProcessData process = (ProcessData)processes.elementAt(m);
                out.println("<b><font size=5 color=YELLOW>PROCESS => <i>" + process.getName() +
"</i></font></b> <br><br><br>");
                out.println("<b><font size=4>TO-DO ITEMS FOR " + userdata.getUsername().toUpperCase()
+ ":</font></b> <br><br>");
                Vector stages = process.getStages();
                <!-- display to-do items for process in table -->
                for(int i = stages.size()-1; i > -1; i--) {
                  StageData stage = (StageData)stages.elementAt(i);
                  if(stage.getUsers().contains(userdata.getUsername())) {
                    out.println("<tr><td>" + stage.getName() + "</td><td>" + stage.getDescription() +
"</td><td>" + stage.getArtefact() + "</td><td>" + stage.getState() + "</td>");
                  }
                }
              }%>
            </table>
          </td>
        </tr>
      </table>
    </div>
    <% }
  </body>
</HTML>
```

Figure 5-19. The Core of ToDoList.jsp.

5.4. Summary

The implementation of the IMÅL prototype process system sees several components and services work collectively to provide users with ongoing process support. Process modelling and enactment functionality are provided via the existing Pounamu software tool, onto which a component to facilitate efficient communication with the process engine was added. The process engine itself was realised as a collection of Web Services with individual responsibilities that together provide all the functionalities crucial to a process enactment engine. This includes a JSP-based to-do list service that allows certified users to obtain lists of to-do items as well as overall process information.

Chapter 6 - Evaluation

6.1. Introduction

This chapter presents an evaluation that was carried out on the IMÅL prototype system
described in the preceding chapters. In the first part of the chapter, a Cognitive
Dimensions evaluation of the system and the process modelling language is proposed.
The second part presents and analyses an evaluation survey that was carried out on the
prototype system.

6.2. Cognitive Dimensions Evaluation

The Cognitive Dimensions (CD) framework was first introduced by Thomas Green in
1989 [37]. It provides a set of discussion tools for evaluating notations and information
artefacts. The framework was created in order to "assist the designers of notational
systems and information artefacts to evaluate their designs with respect to the impact that
they will have on the users of those designs" [11]. Table 6-1 lists and briefly describes the
14 dimensions currently included in the framework.

There are several advantages of using CD as a basis for discussing and assessing features
of a notation or tool before conducting tests involving real users. Firstly, it saves time and
decreases costs by allowing the developers themselves to identify weaknesses and
loopholes that can be fixed before expensive testing procedures are employed.
Additionally, a higher level of dialogue and discussion between developers is prompted,

and the dimensions serve as a basis for informed critique [38]. They can also help clarify previously vague notions and create new goals and ambitions [38].

Table 6-1. The 14 Cognitive Dimensions (after [38]).

DIMENSION	THUMBNAIL DESCRIPTION
Viscosity	Resistance to change
Hidden dependencies	Important links between entities are not visible
Visibility and juxtaposibility	Ability to view components easily
Imposed lookahead	Constraints on order of doing things
Secondary notation	Extra information in means other than program syntax
Closeness of mapping	Representation maps to the domain it intends to describes
Progressive evaluation	Ability to check parts while others are incomplete
Hard mental operations	Operations that tax working memory
Diffuseness/terseness	Succinctness of language
Abstraction gradient	Amount of abstraction required versus amount possible
Role-expressiveness	Purpose of a component is readily inferred
Error-proneness	Syntax provokes slips
Perceptual mapping	Important meanings conveyed by position, size, colour etc
Consistency	Similar semantics expressed in similar syntax

In the following sections, the IMÅL prototype system is evaluated according to the 14 dimensions encompassed by the CD framework.

6.2.1. Viscosity

Viscosity refers to how much effort is required in order to accomplish a goal [11]. There are two different types of viscosity, namely repetition and knock-on viscosity. Repetition viscosity addresses a series of actions of the same type, whereas knock-on viscosity deals with the 'domino-effect' where several additional changes are needed to restore consistency after one change is made [11]. An example of repetition viscosity is when a user wants to change all headings in a document to a larger font size, and has to go through the whole document and change every single heading one by one.

IMÅL generally provides an easy-to-use interface, with which desired goals can be achieved relatively easily. Adding, deleting and updating shapes and connectors, as well as enacting stages and viewing to-do list information, are all examples of tasks that are simple for the user to perform.

However, the fact that the focus in this thesis project was mainly on the process enactment engine of IMÅL resulted in less time being spent on the process modelling part of the system. This has led to limited flexibility in the latter, and a certain degree of knock-on viscosity is therefore evident in this part of the system. For example, if the user wishes to change the output values of a stage, the input values of the following stage must be changed accordingly, in order for the system to work properly. Similarly, if the user alters the name of a role, these alterations have to be manually mirrored throughout the whole process definition.

As do most visual languages and tools, IMÅL suffers from the need to rearrange objects in a view in order to incorporate new objects that are added. This issue is merely related to the use of the Pounamu tool, and has been partly addressed by allowing multiple views for each model project.

6.2.2. Hidden Dependencies

Hidden dependencies occur when entities in the system are dependent on other entities in the system, and these dependencies are invisible [11]. For example, if one part of the system is being referenced by other components, altering this part may have unexpected consequences for the referencing components [11].

There are no hidden dependencies involved when modelling a process in a single view in IMÅL. However, hidden dependencies are introduced when processes are modelled in multiple views. The system currently requires that the user manually keeps the process model consistent across all the views, including updating references to altered components. Ideally, a mechanism that keeps track of the relationships between all the components in different views should be available. This could, for example, be a parent/child-view mechanism, and should be embedded in future versions of IMÅL.

6.2.3. Visibility and Juxtaposibility

Visibility refers to how visible information is in the system, and how easy it is to get any desired part of the program into view [11]. Juxtaposibility is closely related to visibility, and refers to the ability to display parts of the program side by side for comparison purposes. Visibility and juxtaposibility are important considerations for simple activities like reading information as well as more complex activities like creating and modifying information structures in a system [11].

IMÅL offers reasonably good visibility, as functionality for both re-sizing and scrolling of views is provided by Pounamu. Shown not only in the main modelling view but also in a separate property panel, stage properties are particularly visible. Additionally, the multiple views feature aids visibility, as it enables the user to split into multiple views a chunk of information that is too large for easy comprehension in a single view.

The system does not currently provide zoom functionality or a way to display parts of the program side by side except by using multiple windows. These features are, however, purely Pounamu-related, and will be introduced in newer versions of the tool.

6.2.4. Imposed Lookahead

Imposed lookahead, also referred to as premature commitment, can be explained as constraints on the order of performing actions within a system [11]. In other words, it is the degree to which the system forces the user to perform certain actions, in order to complete a desired task, without having all the necessary information available to perform these actions. A typical example of imposed lookahead is when the system forces a user to choose a certain path or declare certain identifiers too early [11].

There are some constraints on the order of performing actions in IMÅL. Similarly to most other visual tools, a problem is introduced when adding a new stage to a process model, namely where to place it. This becomes a problem in situations where the user does not have a clear mental representation of how the final process model is going to look when he/she starts modelling, and is forced to commit prematurely. However, the system provides functionality to move and re-size stages, in order to reduce the impact of this problem.

The fact that the whole process model must be completely defined before the process engine can be plugged in and enactment started is another constraint in IMÅL that might force the user to perform actions without having all the information necessary to do so. An example is if the user wishes to start enacting the process after modelling the first three stages of a process, but is not yet sure what the remaining stages will be. In this situation, the user is forced to make premature decisions about the remaining stages in order to be able to plug in the process engine and start enactment.

6.2.5. Secondary Notation

A secondary notation is a notation used to convey information other than the formal syntax offered by the system [11]. This is important, because users frequently wish to express themselves informally to enhance the overall value of the information. Examples of the use of secondary notations from programming languages are formatting such as indenting and bracketing as well as comments [11].

There are few opportunities for the use of secondary notations in the IMÅL system. This is mainly due to the nature of the system and what is being modelled. In order to enable the enactment of process models they need to be strictly structured, and there is therefore little room for individual expression. Users are, however, able to include their own description of stages, and the layout of the process model is also entirely up to the user. The system also provides some degree of freedom via the multiple views feature that allows users to split processes into smaller, logically related process parts. An example of a way to provide users with secondary means to convey information is to include additional shapes for users' individual notes and annotations.

6.2.6. Closeness of Mapping

Closeness of mapping refers to how closely the system notation relates to the real-world domain or object that it is intended to describe [11]. In other words, it captures the

137

similarity between the domain described and the representation used to describe it. Ideally, the notation should be clearly related to the real-world domain in terms of entities and functionality.

The programming language used in IMÅL is relatively closely related to the software process area. It contains means to model all the basic components in a process, including stage and work information, role-actor associations, and process flow. However, because a certain level of structure is essential in order to achieve an executable process model, the process modelling language is less flexible than a real-world process. The time constraints involved in this project have also been a limiting factor for the notation. For example, the user is currently only able to model one tool and one artefact for each stage, whereas multiple tools and artefacts would typically be needed for a stage in a real-world scenario.

Furthermore, the enactment facilities provided by the process enactment engine are very similar to those performed in the real-world execution of software processes. As in manual software processes, users are able to start, pause, resume and finish work on stages.

6.2.7. Progressive Evaluation

Progressive evaluation can be seen as the level of support for evaluation of work in progress, meaning the ability to stop part-way through development and evaluate the work done so far [11]. The dimension captures how well the system supports the checking of parts completed while other parts are still untouched or under development. An example of using progressive evaluation is when the login functionality of a computer program is developed and tested before any other functions are completed.

IMÅL does not support progressive evaluation of process models in terms of testing incomplete models, as the process enactment engine requires a fully defined process model in order to present the user with the appropriate enactment facilities. Providing support for progressive evaluation is closely linked with supporting evolution of processes and process models. The system should ideally incorporate functionality to deal with both incomplete and changing process models in a way that does not constrain the use of the enactment facilities.

6.2.8. Hard Mental Operations

Hard mental operations refer to situations in which a notation or tool is so complex that using it requires a significant amount of cognitive resources [11]. In other words, hard mental operations make it difficult for the user to complete tasks or understand concepts in the notation or tool. Examples of hard mental operations are combinations of several complex concepts within one task and deeply nested structures [11]. An important issue to take into account when talking about hard mental operations is the fact that some concepts are inherently hard to understand, making the tool or notation that deals with them correspondingly difficult to work with.

Once familiarisation with processes in general is acquired, using IMÅL does not involve hard mental operations. Hence, IMÅL does not embed any complex concepts that are too hard for users to grasp. The process modelling language and tool and the process enactment functionality are all straightforward to understand and easy to use. Basically, a brief introduction or tutorial is enough to teach any process-aware person how to use the system.

6.2.9. Diffuseness/Terseness

Diffuseness is defined as the verbosity of a notation [11]. A diffuse notation uses numerous symbols and a considerable amount of space to convey information. Conversely, terse notations use very few symbols and less space to express similar concepts. Notably diffuse notations can be annoying to work with, difficult to remember, and can decrease the available working area. Notations that are too terse, on the other hand, can be hard to grasp and easily cause misunderstandings resulting from limited expression opportunities. Ideally, a notation should fall to somewhere in between these two extremes.

IMÅL's process modelling language is reasonably simple, and provides only the most basic elements and attributes needed to model a process. The notation does therefore not employ an excessive amount of symbols, yet enough to convey essential information. Additionally, the notation is very flexible when it comes to the amount of space required. This is because Pounamu already provides functionality for re-sizing of objects in a view. Hence, all the elements contained in the process modelling language can be minimised to take up less space, or enlarged to occupy more space, depending on users' preferences. In summary, the process modelling language is neither too diffuse nor too terse, but striking a good balance.

6.2.10. Abstraction Gradient

Abstraction gradient captures the availability of abstraction or redefinition mechanisms in the notation or tool [11]. This dimension is tricky, as providing a certain level of high-level abstraction is beneficial, whereas the inclusion of too many abstractions can lead to a steep learning curve for users. Hence, a desired solution is a tool that provides abstractions to an extent that is understandable by the user. An example of abstraction in a computer program is the use of short-cut keys to initialise the same actions that are activated when choosing menu items with the mouse.

Although not optimal, a certain degree of abstraction is available in IMÅL's process modelling language. For example, the language offers functionality to abstract the notion of human user from the role the user plays in a process. Instead of assigning a task to a person, the task is assigned to a role, and people filling that role are responsible for the completion of the task. An important advantage of this abstraction is that it enables a user to play multiple roles, as well as a role to be filled by multiple users, which is often the case in a real-world software process scenario.

In some situations where it would be useful, IMÅL does not currently support abstraction. For example, as mentioned previously, users are only able to associate one tool with each stage when modelling processes. An abstraction mechanism that allows users to associate collections of tools with a stage could be implemented to improve this. Similarly, in order to conform to a real-world process scenario, users should also be able to model multiple artefacts for each stage in a process.

6.2.11. Role-Expressiveness

Role-expressiveness refers to how easy it is for users to discover and understand the role of each component in a notation or system [37]. A notation that clearly displays its structure is said to be role-expressive, whereas a diffuse notation with little separation of concepts and entities is not.

IMÅL consists of a number of components that are easily distinguishable, and therefore offer relatively good role-expressiveness. On the system level, IMÅL is made up of a process modelling language, a process modelling tool, and a process enactment engine. The role of each of these three components is intuitive and easy to understand. On the process modelling level, the fact that the language provided is reasonably simple means that there are few modelling components to learn and get used to. Stages and role-actor associations are represented by different shapes, and process flow is represented by connectors between stages. Because of a clear visual distinction, it is easy for the user to understand the role of the respective components.

6.2.12. Error-Proneness

To what degree the system encourages mistakes, and fails to satisfactorily protect the user from making them, is determined by the error-proneness of the system [11]. As far as feasible, the system should not allow the user to make mistakes. However, some errors are hard to prevent directly, such as mismatched brackets and spelling mistakes made when using a programming language. In such cases, the system should notify the user about when and where a mistake was made and suggest appropriate corrections.

The process modelling functionality of IMÅL is relatively susceptible to errors, whereas the process enactment part does not allow the user to make mistakes at all. This is a direct outcome of the decision that was made at the start of the prototype development to focus mostly on the process enactment engine and its functionality. The process enactment part only presents possible actions to the users, preventing them from doing anything that will not work or generate errors. After every action taken by the user, the enactment functionality is updated accordingly.

An example of a mistake invited by the process modelling part of the system is the failure to give unique identifiers to each stage in a process. The responsibility of unique identifiers is left entirely in the hands of the modeller, and the process model will not work properly unless this convention is abided by. Ideally, the system should have a built-in mechanism that protects the user from giving two stages the same identifier. Furthermore, spelling mistakes are easy to make when modelling processes, and these may also have a negative impact on enactment of the resulting process model. Adding spell-checking procedures and the ability to reuse expressions can help solve this problem.

6.2.13. Perceptual Mapping

Perceptual mapping refers to the availability of visual cues to convey important information within the notation or tool [38]. The position, size and colour of entities are examples of such visual cues.

An important application of perceptual cues in IMÅL is the use of colours to convey information about process state during process enactment. This enhances the usability of the system by making it very easy for users to see where the process is at, what stages are enacted, and what actions should be performed next. Without these colour cues it would be very hard for users to keep up to date with the process, and as a result the system usability would have been notably reduced. Additionally, the different elements in the process modelling language are represented with different-sized shapes. For example, start-stages and stop-stages are quite small, whereas work-stages are larger, because they are the more essential components.

6.2.14. Consistency

By consistency is meant how consistently the notation and system expresses similar semantics and concepts [11]. An inconsistent system that presents the user with quite different ways to handle similar situations, or achieve similar goals, might suffer from decreased usability. How long it takes for a beginner to infer one part of the notation or system from another is an example of a measure of consistency.

IMÅL offers a high level of consistency in both the process modelling phase and the enactment phase. Once the user has modelled one stage, it is easy for him/her to understand how to model additional stages as well as role-actor associations. These are basically modelled in exactly the same way, apart from the different attributes they require values for. Modelling process flow is also simple, and the users need only be taught how to do one in order to be able to model all the flows in a process.

The process enactment engine also presents similar functionality in a consistent manner. For example, when a user has learnt how to start a stage, he/she can easily infer how to pause, resume and finish stages. In general, a user who has been introduced to the system once will find it fairly consistent and easy to operate.

6.3. Survey

Obtaining feedback from real users regarding both the functionality and usability of the IMÅL prototype served as the main objective for the survey that was undertaken. This section explains the nature of the survey and presents the statistical results obtained from it. Some suggestions for improvement identified by the survey are also discussed, along with their implications for further enhancement of the system. Prior to recruiting participants and going through with the survey, approval was sought from the University of Auckland Human Participants Ethics Committee (UAHSEC). Once approval was obtained, the survey was carried out in the Computer Science Department in conformance with the UAHSEC guidelines [91].

6.3.1. Participants

Participants for the survey were recruited through verbal invitation mainly from inside the University of Auckland. The only requirement for taking part in the survey was having basic experience with visual modelling tools.

143

The survey was completed by a total number of eight participants, of which four had previous experience with the Pounamu meta-CASE tool. A conscious decision was made to include participants both with and without Pounamu experience in order to determine the impact such experience had on how the prototype system was perceived.

6.3.2. Nature of Survey

The survey consisted of two tasks for the voluntary subjects to complete. Firstly, the subjects were asked to individually work through a tutorial using the IMÅL prototype system. Then, they were to answer a questionnaire that focused on their experiences in using the system. A brief introduction to the prototype system was given at the start of the tutorial, followed by in-depth explanations of each task the subjects were to perform. No additional information or instructions were therefore needed to complete the tutorial. The subjects were, however, able to ask questions if they encountered problems or had any queries along the way.

The tutorial and the questionnaire were targeted mainly at the following three activities:

1. **Process modelling** – with IMÅL
2. **Process enactment** – with IMÅL
3. **View to-do list** – in a Web browser

Firstly, the participants were asked to open a new model project using the Pounamu Process Modelling Tool (PPMT). They were then to model a simple software process by adding a start-stage, a stop-stage, three work-stages, three role-actor associations and five flows to the model view. The tutorial further asked the subjects to plug in the process engine using the model project menu bar. When the process engine was plugged in, they were to enact the process model by enacting, pausing, resuming and finishing stages via the stages' drop-down menus. Using the to-do list service was the last activity included in the tutorial. The participants were asked to supply a given username and password to gain access to the service, and click a button to view their to-do list for the process.

If they wished, the subjects were free to explore the tool further. They could model other processes in new model projects, plug in process engines for the new projects, and enact several processes simultaneously. This would also cause the to-do list service to be updated with information about the newly enacted processes.

Finally, the participants were given the questionnaire, and asked to answer all the questions based on their use of the tool. Both open-ended and closed-ended questions relating to each of the three categories above were included in the questionnaire. Gathering individual views and thoughts about the prototype was the objective of the open-ended questions, whereas the rating questions were designed in a way that enabled the answers to be assembled and analysed statistically.

The rating questions for the process modelling part were concerned with how intuitive and easy to use the process modelling language and tool were, and an open-ended question allowed the users to include any additional comments they had about it. Rating questions for process enactment asked how easy the functionality was to use, how satisfactory the process guidance provided was, and how useful the colour changes were in conveying process state changes. As for process modelling, an open-ended question for further comments was included also for the process enactment part. How good and relevant the to-do list service was made up the rating questions for this service, before an open-ended question again allowed for additional comments.

Some general questions were included at the end of the questionnaire in order to get a feel for the overall perception of the system, and to provide openings for general comments and suggestions for improvement. The subjects were asked whether they thought the tool was useful in supporting processes and whether they would consider using such a process support system for their own company. They were also asked if they had any suggestions for improvement of the system, or any other comments, positive or negative, about it. Full details of the tutorial and questionnaire are shown in Appendices A and B respectively.

6.3.3. Statistical Results

An immediate examination of the answers gathered showed that there were no significant differences between the answers obtained from the participants who had previous experience with the Pounamu tool and those that did not. The result set as a whole is therefore examined here, with no reference to the level of acquaintance with Pounamu. Presented and discussed in the following are the results of the rating questions for the modelling, enactment and to-do list functionality.

In general, the prototype system was regarded favourably by the survey participants. Table 6-2 shows all participants' ratings of each individual question, as well as their average rating of each category, namely modelling, enactment and to-do list. Total average ratings and standard deviations for the respective questions and categories are also shown. The rating range was 1-5 where '1' was very easy/intuitive/good, '3' was average and '5' was not at all easy/intuitive/good.

Table 6-2. Results Obtained from Rating Questions (MED – Median, STDEV – Standard Deviation).

PARTICIPANT #	1	2	3	4	5	6	7	8	MEAN	MED	STDEV
Process modelling:											
Easy	3	1	1	2	2	2	1	2	1.75	2	0.661
Intuitive	1	2	1	1	2	2	2	2	1.63	2	0.484
Average for modelling	*2.00*	*1.50*	*1.00*	*1.50*	*2.00*	*2.00*	*1.50*	*2.00*	*1.69*	*2.00*	0.348
Process enactment:											
Easy	1	2	1	2	1	1	2	1	1.38	1	0.484
Guidance satisfactory	1	2	1	2	1	1	1	3	1.50	1	0.707
Colour change useful	1	2	1	1	1	1	1	1	1.13	1	0.331
Average for enactment	*1.00*	*2.00*	*1.00*	*1.67*	*1.00*	*1.00*	*1.33*	*1.67*	*1.33*	*1.00*	0.373
To-do list:											
Provided info well	3	2	1	2	2	2	1	1	1.75	2	0.661
Relevant	1	2	1	1	1	1	1	1	1.13	1	0.331
Average for to-do list	*2.00*	*2.00*	*1.00*	*1.50*	*1.50*	*1.50*	*1.00*	*1.00*	*1.44*	*1.50*	0.39

Figure 6-1 depicts the average score for each rating question graphically, whereas the histogram in Figure 6-2 shows the average rating score within each category.

Process modelling was the part with the worst average rating out of the three categories (mean – 1.69; median – 2; standard deviation – 0.348). This does not mean, however, that the ratings obtained for the process modelling functionality were not good. On average, all the participants thought the process modelling language and tool were intuitive and easy to use. The fact that the process modelling language got the lowest rating is not surprising, as the process modelling category is the one in which the learning curve is steepest. This is because the user has to learn a new language, as well as the tool that implements this language. Using the process enactment and to-do list functionalities requires considerably less effort from the user in terms of learning.

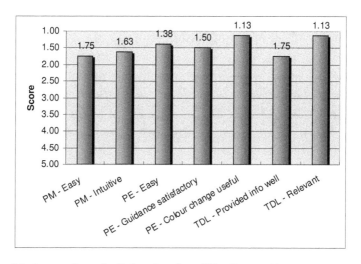

Figure 6-1. Average Score for Rating Questions (PM – Process Modelling, PE – Process Enactment, TDL – To-Do List).

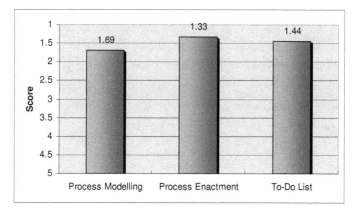

Figure 6-2. Average Score for each Category.

147

Process enactment received the best average rating out of the three activities (mean – 1.33; median – 1; standard deviation – 0.373). In general, the participants found the process enactment functionality easy to use and the guidance on process status satisfactory. All the participants except one perceived the changes in stage colour according to process enactment as very useful. As the main focus of the prototype development undertaken in this thesis project was on the process enactment engine, the overall ratings obtained for the process enactment functionality were very gratifying.

The view to-do list functionality was also rated favourably by the participants of the survey (mean – 1.44; median – 1.5; standard deviation – 0.39). They generally believed the service provided information about the progress of the process well, and everyone except one answered that they believed providing such a service was very relevant.

As mentioned earlier, some more general questions at the end addressed the prototype system as a whole rather than any specific functionality. All the subjects answered yes to the question about whether they thought the prototype was useful in supporting processes. Similarly, to the question about whether they would consider using such a tool in their office environment, a positive response was obtained from all participants.

6.3.4. Suggestions for Improvement

This section presents and discusses the feedback obtained from the open-ended questions included in the questionnaire. As mentioned previously, these questions requested suggestions for improvement of the system, as well as general positive or negative comments about the system. They were included in order to capture any additional thoughts the participants had on the system after using it to complete the tutorial, and to identify what improvements users of the system would prefer.

The participants had several positive comments about the tool and its features, such as "easy and intuitive to use", "good basic functionality for process modelling", "process enactment functionality very easy to use and changes highly visible" and "to-do list service very useful at showing process state". However, the focus in the following discussion is on features the participants thought were missing and suggestions for improving the system.

One participant stated that only allowing the user to model one tool and one artefact per stage is a limitation of the modelling functionality. This is an important observation, and a reflection of the fact that the focus in this prototype was on the enactment engine, which led to less time being spent on the process modelling language and tool. Ideally, the user should be able to model multiple artefacts and tools per stage to better reflect a real-world process scenario.

A complaint about the to-do list service was that the user has to scroll down every time the page is refreshed in order to see the full list of processes and stages. The feasibility of improving this depends, of course, on how much information needs to be displayed (how many processes are being enacted), as the amount of information a Web browser accommodates without scrolling is limited. However, it is certainly possible to utilise better the space immediately available in the Web browser in order to reduce the amount of scrolling needed.

Allowing users to enact processes via the to-do list service was another suggestion for improvement. This would be a very useful enhancement that would make the enactment functionality more easily accessible, as it would mean that all user interaction in terms of process enactment can be handled entirely by a Web browser. Hence, users would not necessarily need to have a copy of Pounamu installed on their computers in order to use the IMÅL system.

The remaining suggestions for improvement involve all the different parts of the system, and fall into the five categories: stage automation, tool integration, process evolution, time concepts, and user-types.

The possibility to automate stages and have the system make automated decisions without user intervention was suggested by several participants as a way of considerably enhancing the tool. Such automation could be achieved by adding an automation component that takes care of and processes all stages that has an 'automation' flag set to true.

Additionally, integration of third-party tools was seen as an important addition to the system, in order to support users in performing work on stages throughout processes. A way of doing this is by having the system automatically starting up the tool and artefact needed when a user enacts a stage.

One topic mentioned was that the system could be improved in such a way that it would be able to deal with changing process models. Ideally, the system should support evolution of processes by effectively handling changes to process models both before, during, and possibly also after, process enactment. It might also be beneficial to offer support for enactment of incomplete process models, which would enable people to start work on already defined stages of a process while other stages are still being modelled.

Another participant suggested the incorporation of time concepts into the system, for example, in terms of tracking the time spent on and adding deadlines for stages. Such functionality would have to first be incorporated at the modelling level, and then followed up at the enactment level. A way in which this could be achieved is by using the already existing functionality offered by Pounamu's event handlers. These event handlers are user-defined entities that determine actions to be taken when certain events occur, and could be programmed to deal with time-tracking issues. Furthermore, the to-do list service could display time-related information along with the other information currently shown for each stage. Examples of time-related information that could beneficially be shown in the to-do list are how much time has been spent on a stage, what the estimated time for the stage is, and whether or not it is overdue.

A last suggestion was to integrate into the system the notion of user-types with different rights when it comes to using the system. This is merely an abstraction issue, and would improve the overall applicability of the system. A user-type is a status that can be assigned to users, and it is associated with rights to perform actions within the system. As with time-tracking, user-types need to be modelled in the process modelling phase, and accordingly supported by the process enactment engine. At present, IMÅL allows for modelling of such user-types, but the enactment engine is not implemented to deal with the different types. For example, a user filling the role 'manager' could be allocated administrative rights for process modelling and enactment, as well as rights to view information about all users' stages in the to-do list service. User role 'employee', on the other hand, might only be allowed to enact and view the stages for which he/she is responsible.

6.4. Summary

The CD evaluation proved to be a good tool for evaluating the IMÅL prototype before employing real users in a usability test. Several issues where identified simply by going through each dimension with regard to the prototype. For example, the tool beneficially provides good closeness of mapping, high consistency, and few hard mental operations, however, the modelling functionality is reasonably error-prone. Some of the issues identified, such as insufficient visibility and juxtaposibility, are merely associated with the implementation of the Pounamu software tool.

Feedback obtained from the evaluation survey that was carried out was very positive, and showed that the development so far had been reasonably successful. In particular, the feedback on the usability of the system was positive, and most of the suggestions for improvement addressed possible additional functionality rather than the usability of the existing functionality. Several useful suggestions for future improvement were gathered, which will serve as a basis for further development increments to improve the IMÅL prototype system. Some of these suggestions are addressed in the next chapter of this thesis report, whereas others are left for future work.

Chapter 7 - Integrating Existing External Systems

7.1. Introduction

With the intention of addressing some of the weaknesses identified by the evaluation described in Chapter 6, this chapter describes the integration of some existing tools with the IMÅL system. The third-party tools that were integrated were Microsoft Infopath and Idiom Decision Suite. Infopath was integrated to serve as a coordination service for the flow of artefacts between stages and users. This was done in form of a tool initiation service that automatically starts up Infopath and the relevant artefact when users enact stages with which Infopath is the associated tool. The tool initiation service was also enabled to automatically start up other commonly used documents and programs, such as Microsoft Word, Excel and Visio, JCreator, Adobe Acrobat Reader and Internet Explorer. Idiom was integrated as an automation service, in order to enable automation of stages and tasks that need no user intervention. The automation was accomplished via a rule-based specification in Idiom.

The two new components added to the IMÅL prototype, namely the tool initiation service and the stage automation service, were designed as Web Services to conform to the service-oriented architecture of the IMÅL system. Because of time constraints, they were not implemented as actual Web Services but hard-coded locally. They could both, however, easily be adjusted to work in a Web Services environment simply by wrapping them in Web Services code and deploying them on remote servers.

This chapter presents the requirements identified for the integration of Infopath and Idiom, as well as the design and implementation of the corresponding components. Furthermore, an informal evaluation of the integration is proposed. Towards the end of the chapter, a section explains some additional improvements that were made to the to-do list service on the basis of the feedback gathered from the survey presented in Section 6.3.

7.2. Motivation

To improve the overall system by addressing some of the deficiencies identified in Chapter 6 served as the main motivation for integrating existing third-party tools with the IMÅL system. The reasons why Infopath and Idiom were chosen for these purposes are expanded on in the following.

Infopath is a new member of the Microsoft Office family as of 2003. It provides means to create and work with rich and dynamic forms, in order to enable teams and organisations to easily gather and share information across a wide range of formal and informal business processes [59].

The Infopath software program fits well into this thesis project, as it provides access to documents at remote locations via Web Services. As the documents created using Infopath are XML-based, they can easily be transported as SOAP attachments to and from a Web Service at a remote location, and stored in a document repository. Infopath therefore presents an ideal way in which documents associated with a process can be retrieved, updated, stored and directed between different stages and users.

Idiom Decision Suite is described as a tool that "captures the specification of the business decisions that support core business processes, and then deploys those decisions to any required system, including third party systems" [46]. Basically, the tool makes decisions involving several variables on the basis of a pre-defined specification. The specification is defined by a domain expert using the Idiom Decision Manager before the Idiom Generator turns it into code that can be plugged into third-party tools.

There were several reasons for choosing to integrate Idiom with IMÅL. Firstly, the fact that Idiom is developed with the intention of plugging it into third-party systems made the integration feasible. Secondly, Idiom enables the triggering of events depending on the decision outcomes of processing information. This is extremely useful in a process support system, because not only does it enable automatic decision-making, but also the automation of events caused by the decision outcome. Another reason for integrating Idiom was that the functionality can be made available as a Web Service and deployed remotely, which conforms perfectly to the architecture of the IMÅL system.

Idiom could also be used in the implementation of a complex flow component, which is another component needed to further advance the process support system. Idiom would be ideal for the development of such a component, because complex flow is a scenario where the values of several variables together determine where in the process to proceed to after an event occurs.

7.3. Requirements

The additional functionality described in this chapter addresses two of the general requirements identified in Section 3.2.1, namely the integration of external tools and the automation of process parts. These two requirements are elaborated on in Section 3.2.1.3 and Section 3.2.1.5 respectively, to which the reader is referred for further explanations. Presented here are detailed functional requirements for the two new services, as well as use cases and an OOA diagram.

7.3.1. Functional Requirements

Specific requirements for the two new components that were developed are outlined in the following sub-sections.

7.3.1.1. The Tool Service

The tool initiation service should provide users with the tools and artefacts they need to be able to complete work on stages. This should be provided by automatically opening the tool and document associated with a stage when it is enacted by the user. To activate the

155

service, the user should not need to do anything else but indicate that he/she is starting work on a stage by enacting that stage in the process model.

Furthermore, when the user closes the document, the system should ask the user whether he/she has completed the stage, and act according to the user's response. Hence, if the user responds that the stage is completed, the system should fire a finish-stage event for the related stage.

7.3.1.2. The Automation Service

The overall task of the automation service is to automatically complete tasks without user intervention. An important requirement of this component is that it should work in the background. In other words, the user should not be affected in any way by the processing of automatic tasks other than possibly the outcome of the processing. The stage automation service built in this project serves as a proof of concept, and therefore has limited functionality. It is aimed at processing one specific task highly relevant to process management, namely the monitoring and control of the amount of time spent on processes.

The service should accept XML-formatted information regarding how much time each team member has spent on tasks in a process, and calculate the total number of hours spent for the whole team. Furthermore, the service should compare the resulting number against the number of hours estimated for the process, and send an email notification to the project manager if the total number of hours spent is more than 10% above the estimate.

7.3.2. Use Cases

Figure 7-1 shows two use cases that capture the use of the new functionality that was added on to the IMÅL system. The use cases are both refined versions of Use Case 3 described in Section 3.5.3. In the following, the description of the new use cases are supplemented with screen shots of the use case scenarios occurring in the IMÅL system.

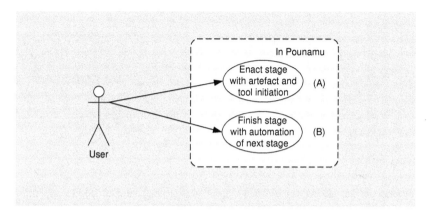

Figure 7-1. Use Case Diagram Capturing the Use of the New Functionality.

7.3.2.1. Use Case A: Enact Stage with Artefact and Tool Initiation

Name: Enact stage with artefact and tool initiation.

Description: Enact a stage leads to the appropriate tool and artefact being opened.

Actor: All certified users.

Precondition: The process engine component must be plugged in.

Result: The user is automatically presented with the tool and artefact needed to complete the work associated with the stage.

Flow of events:

4. The user chooses the enactment event 'enact stage' from a stage's drop-down menu.

5. Pounamu's process engine component forwards the enactment event to the process engine service.

5.1. The process engine service delegates the enactment event to the appropriate external service for processing.

1.1.5. The external service processes the enactment event, and records in the refresher service updates to the user interface and process model accordingly.

1.2.5. The artefact and tool associated with the stage are returned to the process engine service.

157

5.2. The process engine service returns the artefact and tool to Pounamu's process engine component.

6. Pounamu's process engine component calls the refresher service via the process engine service to obtain necessary updates to the user interface and process model, and applies them in Pounamu.

7. Pounamu's process engine component calls the tool service via the process engine service to retrieve the appropriate document from a document repository and open it with the appropriate tool.

In order to demonstrate the use of the new functionality added to IMÅL, a modified version of the software update scenario used throughout this thesis report is employed in this chapter. The most important changes are that the first stage now has an Infopath document associated with it, and that an automatic stage to check time spent on the process has been added. Figure 7-2 shows an example of Use Case A in two screen shots. The user enacts a stage with which an Infopath document is associated (Screen Shot 1), and IMÅL automatically initialises Infopath and opens the document for the user (Screen Shot 2).

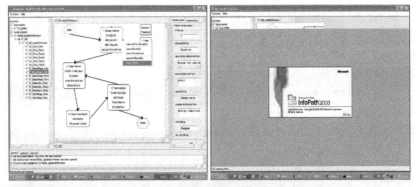

Figure 7-2. Use Case A: Enact Stage with Artefact and Tool Initiation.

7.3.2.2. Use Case B: Finish Stage with Automation of Next Stage

Name: Finish stage with automation of next stage.

Description: The user finishes a stage that leads to an automatic stage, and this stage is automatically completed by the system.

158

Actor: All certified users.

Precondition: The process engine component must be plugged in.

Result: A process stage is completed without user intervention.

Flow of events:

1. The user chooses the enactment event 'finish stage' from a stage's drop-down menu.

2. Pounamu's process engine component forwards the enactment event to the process engine service.

2.1. The process engine service delegates the enactment event to the appropriate external service for processing.

1.1.2. The external service processes the enactment event, and records in the refresher service updates to the user interface and process model accordingly.

2.2. The process engine service checks whether the next stage to be enacted is an automated stage. If it is, the process engine delegates the next stage to the automation service. If the next stage is not an automated stage, go to 3.

2.1.2. The automation service executes the work associated with the stage and records updates to the user interface and process model in the refresher service accordingly.

3. Pounamu's process engine component calls the refresher service via the process engine service to obtain necessary updates to the user interface and process model, and applies them in Pounamu.

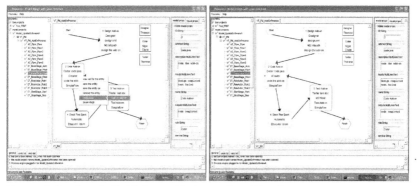

Figure 7-3. Use Case B: Finish Stage with Automation of Next Stage.

159

Figure 7-3 illustrates Use Case B in two screen shots. It starts with the user finishing work on a stage that is followed by an automatic stage (Screen Shot 1). The process engine detects this, automatically processes the next stage, and updates the user interface and process model accordingly (Screen Shot 2).

7.3.3. OOA Class Diagram

Figure 7-4 depicts the two new components in an OOA class diagram. As described in Section 3.6.2, class ToolService provides functionality to automatically activate third-party tools needed to perform work on a stage, whereas class AutomationService is responsible for automation of tasks.

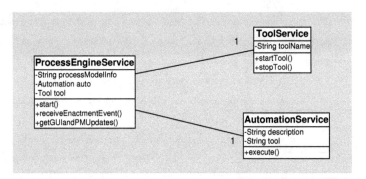

Figure 7-4. OOA Class Diagram for the New Tool Integration Services in the Process Engine Service.

7.4. Design

The components outlined in this chapter are included in the general architectural design diagram presented in Figure 4.2. They are both Web Services that can be deployed at any location and used by the main process engine component. In this section, two OOD class diagrams present refinements of the OOA diagram shown in Figure 7-4. Furthermore, sequence diagrams are included, showing essential event flows that occur when the components are being used by the system. These sequence diagrams are extensions of the

160

sequence diagram presented in Figure 4.13, which captures event flows in an enactment situation. Supplemented with explanations, the class and sequence diagrams shown here capture the detailed design of the tool and automation components.

7.4.1. OOD Class and Sequence Diagrams for the Tool Service

Like the other services in the process engine, described in Section 4.3.2, the core of the tool service is made up of an interface, ToolServiceIF, and an implementation of this interface, namely class ToolServiceImpl. Class ToolServiceImpl additionally accesses a document repository, in which process-related documents are stored.

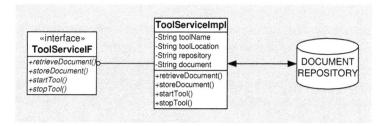

Figure 7-5. OOD Class Diagram for the Tool Service.

As shown in Figure 7-5, the overall design of the tool service is similar to that of the database services, described in Section 4.3.2.1, with an important difference being that it deals with the storing and retrieving of whole documents rather than core data. Hence, the functionality offered by the tool service is slightly different from that of the database services. Class ToolServiceIF declares methods for retrieving and storing documents, as well as opening and closing documents and tools, while class ToolServiceImpl provides the full implementation of the corresponding methods.

The sequence diagram in Figure 7-6 illustrates the main event flows that occur in an enactment situation in which the tool service is utilised. This situation is initiated by the user enacting a stage in a pre-defined process model in Pounamu (1). The enactment event is forwarded through to the simple flow service, which processes the stage and performs necessary updates (2-7). Further, Pounamu's process engine component calls

161

the tool service via the process engine service to initialise the appropriate tool and artefact (8-9). Finally, the process engine component retrieves the updates from the refresher service via the process engine service (10-11), and applies them in Pounamu (12-13).

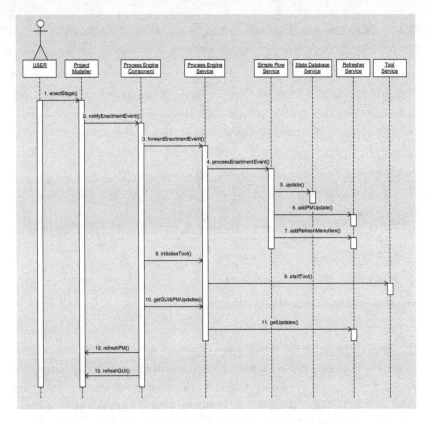

Figure 7-6. Sequence Diagram of a Situation in which the Tool Service is Used.

7.4.2. OOD Class and Sequence Diagrams for the Automation Service

Figure 7-7 displays an OOD diagram capturing the design of the automation service. The core of the service consists of interface IdiomIF and its implementation class IdiomImpl, which declares and implements the method that automatically processes a stage. To process automatic stages, class IdiomImpl utilises class LocalDecisionServer, which is a class provided by Idiom Decision Suite. When processing stages, class LocalDecision-

162

Server employs a collection of classes generated by Idiom on the basis of a pre-defined specification. These classes are plugged in as a part of the service, and because their use is coordinated by class LocalDecisionServer, their underlying implementations need not be known.

When a stage has been processed, the result comes back to class IdiomImpl, and an email notification is sent to the project manager if necessary. This is done by class Email, which contains information about the receiver and sender of the message, the message to be sent and a method to perform the sending.

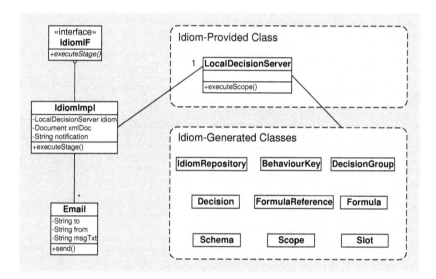

Figure 7-7. OOD Class Diagram for the Automation Service.

Shown in Figure 7-8 is a sequence diagram of the main event flows in a situation where a user finishes work on a stage, followed by the automation service automatically processing the next stage. Firstly, the user finishes a stage in a process model in Pounamu (1). As usual, the enactment event is forwarded through to the simple flow service, which processes the stage and performs necessary updates (2-7). The process engine service then checks whether the next stage is automatic (8), and if it is, forwards it to the automation service for processing (9). Further, the automation service makes use of its Idiom component to process given information about time spent on the process (10), and sends

163

an email to the project manager if the total time spent is over 10% more than estimated (11). Finally, the process engine component retrieves the updates from the refresher service via the process engine service (12-13), and applies them in Pounamu (14-15).

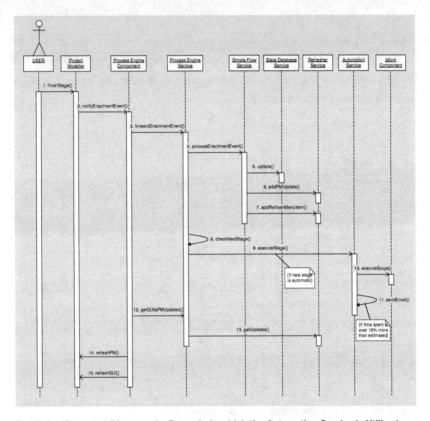

Figure 7-8. Sequence Diagram of a Scenario in which the Automation Service is Utilised.

7.5. Implementation

This section presents implementation details for the tool and automation services, and includes code samples and screen shots with explanations. The external technologies used, mainly Microsoft Infopath and Idiom Decision Suite, were introduced in Section 7.2, and the focus here is on how they were integrated into the IMÅL system.

7.5.1. Tool Start-Up with Microsoft Infopath

The tool service consists of interface ToolServiceIF, which declares methods for starting and stopping external tools, and its implementation, class ToolServiceImpl, which provides the detailed code for the corresponding methods. Class ToolServiceImpl contains a pointer to the document repository location, as well as pointers to the executable files for the external tools that are to be initialised by the service.

As it provides facilities to store and retrieve documents remotely via Web Services, Microsoft Infopath was the main focus of the tool integration enabled in IMÅL. Because of time constraints, the repository for the prototype system was developed locally on a user's computer. However, once a remote repository Web Service to deal with the documents is set up, the only essential change needed in order to use the remote repository rather than the local one, is to set the location of the repository to the URL (Uniform Resource Locator [95]) representing the address of the Web Service. Altering the repository to a Web Service of its own would make the tool service conform strongly to the Web Services oriented architecture of the IMÅL system.

```
/** ToolServiceImpl.java: responsible for tool and document initialisation */
public class ToolServiceImpl {
    String repository = "C:/pounamu/processengine/repository/";
    String infopath = " C:/Program Files/Microsoft Office/Office11/INFOPATH.EXE";
    Process tool = null;

    /** constructor */
    public ToolServiceImpl() {}

    /** open a document using the appropriate tool */
    public void startTool(String document) {
        // get current runtime
        Runtime runtime = Runtime.getRuntime();
        if(document != null && ! document.equals("")) {
            try {
                if(document.endsWith("xml")) {
                    // open document and tool
                    document = infopath + " " + repository + document;
                    tool = runtime.exec(document);
                }
            }
            catch(Exception e) {
                System.out.println("Exception in starting tool: " + e.toString());
            }
        }
    }
}
```

Figure 7-9. The *startTool* Method of Class ToolServiceImpl.

165

The actual starting up of Infopath is performed by the *startTool* method, simplified and shown in Figure 7-9, which takes as a parameter the document that is to be opened. It obtains the current runtime, retrieves the appropriate document from the repository, executes the executable file for Infopath, and opens the document. This event is depicted in two screen shots in Figure 7-2. In future implementations to improve the IMÅL system, the initialising of third-party tools on users' computers could be achieved simply via special Web Services messages. The messages would need to contain the required information, and on that basis initiate the appropriate actions for opening the desired tools and artefacts.

Automatic start-up of several other relevant external tools was also enabled in IMÅL. Documents in formats such as Microsoft Word, Excel and Visio, JCreator, Adobe Acrobat Reader, and Internet Explorer, can all be automatically opened if they are stored locally on a user's computer or on the same internal network. An issue arising, however, is that the storing and retrieving of these document formats to and from a remote document repository is not as easily achievable as with Infopath, which has these facilities built in. Alternatives include manually building a separate repository for these other document formats, or adding an intermediate step that converts the respective documents to and from Infopath format before using the facility provided by Infopath. Texcel's FormBridge is an example of a tool that can be used to covert documents into Infopath format [90]. Ideally, the user should be able to attach any document to an Infopath form and utilise Infopath's facilities.

The *stopTool* method in class ToolServiceImpl defines what happens when the user closes a document that was originally opened by IMÅL. It pops up a dialogue box asking if the user has finished the work associated with the corresponding stage. If the user responds that the stage is completed, a finish-stage event for this stage is fired automatically and consequently handled like any other such event.

7.5.2. Stage Automation with Idiom Decision Suite

Interface IdiomIF along with its implementation class IdiomImpl are the most important elements of the automation service. Figure 7-10 shows the core of class IdiomImpl, and

```java
/** IdiomImpl.java: responsible for processing automatic stages */

public class IdiomImpl {

  String xmlFile = "Etrue.xml";
  String timeReport = "TimeReport.xml";
  String notify = "";
  String notification = "";

  /** constructor */
  public IdiomImpl() {}

  /** execute an automatic stage */
  public void executeStage() {
    try {
      // set up document builder and parse XML file
      DocumentBuilderFactory dbf = DocumentBuilderFactory.newInstance();
      DocumentBuilder db = dbf.newDocumentBuilder();
      Document xmlDoc = db.parse(new File(xmlFile));
      // get an instance of decision server and execute scope
      LocalDecisionServer idiom = LocalDecisionServer.getInstance();
      Vector list = new Vector();
      list.addElement(xmlDoc);
      Date d = new Date();
      list = (Vector)idiom.executeScope(new Date(), list, "Notification", "IDM15");
      // process the result of the execution
      Document result = (Document)list.elementAt(0);
      NodeList nodes = result.getElementsByTagName("Notify");
      notify = (String) nodes.item(0).getFirstChild().getNodeValue();
      // assemble and send email notification to project manager if notify is true
      if(notify.equals(true)) {
        notaification = "Time violation notification: ";
        NodeList nodes1 = result.getElementsByTagName("TimeSpent");
        String timeSpent = (String)nodes1.item(0).getFirstChild().getNodeValue();
        notification += "\n\nYour team has spent " + timeSpent + " hours on this project so far.";
        NodeList nodes2 = result.getElementsByTagName("TimeEstimated");
        String timeEstimated = (String)nodes2.item(0).getFirstChild().getNodeValue();
        notification += "\nThis is over 10% more than the " + timeEstimated + " hours estimated for
the project at this stage.";
        notification += "\n\nFor more information go to: \n\nfile:///" + timeReport;
        String manager = getManagersEmailAddress();
        Email email = new Email(manager, notification);
        email.send();
      }
    }
    catch(Exception e) {
      System.err.println("Exception in executing stage: " + e.toString());
    }
  }
}
```

Figure 7-10. The *executeStage* Method of Class IdiomImpl.

includes the *executeStage* method, which is essential to the enabling of stage automation. Idiom is utilised by the *executeStage* method by making a call to an obtained instance of class LocalDecisionServer in order to automatically execute a stage. Class LocalDecisionServer then uses classes generated by the Idiom Generator to process information and make a decision.

The code produced by the Idiom Generator was generated from a rule specification defined using the Idiom Decision Manager. Idiom Decision Manager refers to the definition of a rule specification as repository administration, and the resulting repository is stored as an XML file [47]. Figure 7-11 shows a screen shot of the repository specification that was defined for use in the automation service. It is associated with an XML Schema that defines the valid format for a business object. A business object is the information supplied to Idiom as the basis for decision-making [47]. The XML Schema, referred to in the repository as 'Notification', is depicted in Figure 7-12.

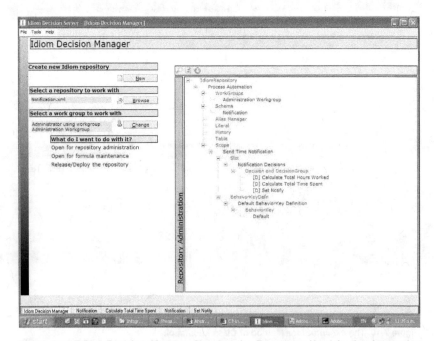

Figure 7-11. Idiom Decision Manager Showing the Repository Used in the Automation Service.

168

```
<!-- ValidBOFormat.xsd: defines the valid format of business objects -->
<?xml version="1.0" encoding="UTF-8"?>
<xsd:schema xmlns:xsd="http://www.w3.org/2001/XMLSchema" elementFormDefault="qualified"
attributeFormDefault="unqualified">
    <!-- root element: TimeNotification -->
    <xsd:element name="TimeNotification">
        <xsd:complexType>
            <xsd:sequence>
                <xsd:element name="Notify" type="xsd:boolean"/>
                <xsd:element name="TimeEstimated" type="xsd:decimal"/>
                <xsd:element name="TimeSpent" type="xsd:decimal"/>
                <!-- multiple actors each with a list of values -->
                <xsd:element name="Actor" minOccurs="0" maxOccurs="unbounded">
                    <xsd:complexType>
                        <xsd:sequence>
                            <xsd:element name="Name" type="xsd:string"/>
                            <xsd:element name="HoursWorked" type="xsd:decimal" minOccurs="0"
maxOccurs="unbounded"/>
                            <xsd:element name="TotalHoursWorked" type="xsd:decimal" minOccurs="0" maxOccurs="1"/>
                        </xsd:sequence>
                    </xsd:complexType>
                </xsd:element>
            </xsd:sequence>
        </xsd:complexType>
    </xsd:element>
</xsd:schema>
```

Figure 7-12. XML Schema Defining the Valid Format for Business Objects.

An example of the information supplied via business objects when making calls to class LocalDecisionServer is shown in Figure 7-13. The structure of the information conforms to the XML Schema shown in Figure 7-12. It is supplied in XML format, as this is the format required by Idiom, and the 'Notify' tag is reserved to hold the result of the processing. This tag is of type boolean, and can therefore only have the values true or false.

```
<!-- BOInfo.xml: example business object information -->
<TimeNotification>
        <Notify></Notify>
        <TimeEstimated>10</TimeEstimated>
        <TimeSpent></TimeSpent>
        <Actor>
                <Name>Nigel</Name>
                <HoursWorked>5.22</HoursWorked>
                <HoursWorked>2.77</HoursWorked>
                <TotalHoursWorked></TotalHoursWorked>
        </Actor>
        <Actor>
                <Name>Claire</Name>
                <HoursWorked>3.02</HoursWorked>
                <TotalHoursWorked></TotalHoursWorked>
        </Actor>
</TimeNotification>
```

Figure 7-13. Example Information Supplied to Idiom for Automatic Stage Completion.

Furthermore, the repository specification shown in Figure 7-11 contains three important decisions, namely 'Calculate total hours worked', 'Calculate total time spent' and 'Set notify'. Seeing that the host environment need only know what decisions are available and not the underlying details of how the decisions are made, these details are hidden. Idiom does this by encapsulating the details within formulas internal to Idiom and invisible to the host application [47].

Creating the formulas and linking them to their corresponding decisions is a part of repository administration. A formula is an operation that consists of a set of operations. It forms a tree of operations in which the root is a single value that represents the outcome of the corresponding decision [47]. As an example of how formulas provide the details for decision-making, the formula that was defined for the 'Set notify' decision used in the automation service is depicted in Figure 7-14. This decision sets the value of the 'Notify' tag to true if the total time spent on a process is over 10% more than the estimate, and false otherwise.

Figure 7-14. The Formula for the 'Set Notify' Decision.

When receiving the call to automatically execute a stage, class LocalDecisionServer coordinates the processing of information and updating of values accordingly, before returning control to class IdiomImpl. Class IdiomImpl then checks the result of the processing by looking up the 'Notify' tag. If it is true, class Email is instantiated and used to automatically send an email notification to the project manager. The core of class Email is shown in Figure 7-15. It contains the message text, the email addresses of the sender and receiver of the message, and the *send* method that performs the actual sending.

```java
/** Email.java: responsible for sending emails */
public class Email {
  String host = "smtp.ec.auckland.ac.nz";
  String to;
  String from = "thel004";
  String msgText;

  /** constructor */
  public Email(String to, String msgText){
    this.msgText = msgText;
    this.to = to;
  }

  /** send an email notification */
  public void send() {
    // create properties and get default session
    Properties props = new Properties();
    props.put("mail.smtp.host", host);
    props.put("mail.debug", "true");
    Session session = Session.getInstance(props, null);
    session.setDebug("true");
    try {
      // create message
      Message msg = new MimeMessage(session);
      msg.setFrom(new InternetAddress(from));
      InternetAddress[] address = {new InternetAddress(to)};
      msg.setRecipients(Message.RecipientType.TO, address);
      msg.setSubject("Important notice! Process " + processID + ": " + processName);
      msg.setSentDate(new Date());
      msg.setText(msgText);
      // send message
      Transport.send(msg);
    }
    catch (MessagingException mex) {
      System.out.println("Exception in sending email: " + mex.toString());
    }
  }
}
```

Figure 7-15. The Core of Class Email.

The two screen shots in Figure 7-3 show the state of the process before (Screen Shot 1) and after (Screen Shot 2) automatic stage processing. Figure 7-16 depicts the automatic email sent by the service to the project manager if the time spent is over 10% more than estimated. A link for the manager to follow in order to view more detailed information is

171

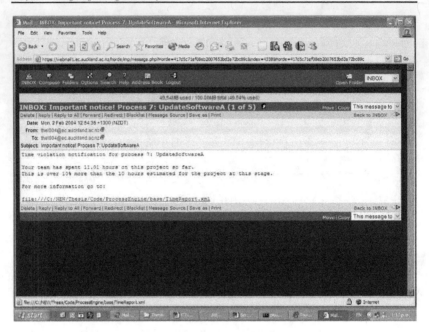

Figure 7-16. An Automatic Email Sent by the Automation Service.

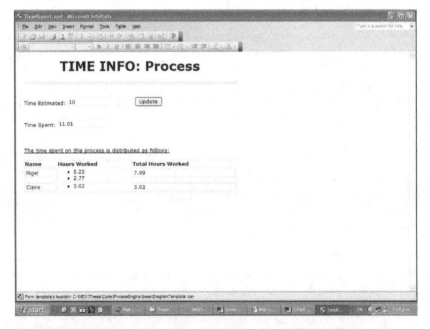

Figure 7-17. Resulting Information from the Automation Service Displayed in Infopath.

included at the bottom of the email. If the manager clicks on this link, more detailed information about the time spent is displayed in Infopath. This is depicted in Figure 7-17, and the information shown is the result of processing the information displayed in Figure 7-13.

7.6. Evaluation

In the following, an informal evaluation is presented of the two tool integration components that were added to the IMÅL prototype system. Neither of the components impacts on the user interface in any way, as all their processing is completed in the background and without user intervention. Hence, usability is not an essential evaluation issue. This section is therefore aimed solely at assessing the additional features provided by the new services in terms of functionality and how useful they are for the IMÅL system.

7.6.1. The Tool Service

The tool service provides good integration of external tools, particularly Microsoft Infopath, and contributes to the user experiencing IMÅL and its integrated external tools as one large environment, rather than several separate tools. It also increases users' efficiency, and makes their lives easier, by presenting them with all the tools and documents they need in order to perform work. The fact that the tool integration is performed automatically is another benefit of the service, as no user intervention is needed, and the user can concentrate solely on his/her work while leaving responsibility for appropriate tool initialisation entirely to IMÅL.

A weakness of the current tool service is that it only deals properly with documents stored locally on a user's computer or on the internal network the computer is on. In order to optimise the tool service to fully conform to IMÅL's Web Services oriented architecture, it needs to be updated to work appropriately with a remote document repository. This would also aid document flow throughout processes, as the remote repository would serve as a kind of document transfer mechanism. Infopath provides facilities to store and retrieve documents to and from remote Web Services, and was therefore chosen as the primary tool for integration. In other words, Infopath's built-in features can be further exploited in order to address current weaknesses of the service. The interface to Infopath needs, however, to be extended to a fully Web Services compatible interface, in order to make for a stable and interoperable integration.

7.6.2. The Automation Service

The automation service also provides integration of an external tool, namely the Idiom Decision Suite. However, the tool integration provided by the automation service does not directly affect the user, but the IMÅL system itself. IMÅL integrates Idiom in order to perform automatic processing on behalf of the system, and thereby enhance the overall functionality of the system and the user's perception of it. Hence, the service is a very powerful addition to the IMÅL system.

Because of the time constraints associated with this thesis project, the scope of the automation service in this implementation was reduced to the handling of one specific scenario, namely checking the time a team has spent working on a process. The service deals very well with this specific scenario, always computing the correct values and triggering the appropriate actions according to the result of the computations. In other words, the functionality provided by the service is optimal for this scenario. Additionally, it invites incorporation of the time dimension into the system in the future.

In order to enable the automation service, and particularly its Idiom component, to handle the scenario described above, a rule specification was defined using the Idiom Decision Manager. This definition process was initially a little cumbersome, as several new concepts had to be learnt and understood in order to use the tool optimally. However, when a good understanding of the tool and its functionality was acquired, the definition of the required rules was straightforward and reasonably effortless. Similarly, plugging the code generated from the specification into the system was troublesome to start off with, but proved to be considerably easier once the steps involved were learnt.

Although the automation service is successful at processing the specific scenario it is

174

designed for, the fact that it can only be applied to this scenario is a disadvantage of the current version of the service. Ideally, the service should be able to handle all sorts of events that can be processed automatically without user intervention. This would require the integration and use of several other components and external tools, which would make the automation service a very substantial and powerful service. Idiom Decision Suite can also be utilised advantageously in future improvements of the automation service, as rule specifications for numerous different scenarios can easily be defined once knowledge about using the tool has been acquired.

7.7. Additional Features in the To-Do List Service

Besides introducing the tool and automation services, some additional features were incorporated into the to-do list service in order to improve the overall IMÅL system.

Firstly, the scrolling problem identified in the survey described in Section 6.3 was addressed. By moving the text further up on the page, as well as making it more compact, the to-do list was modified so that the user can view more information without scrolling. Some scrolling is, however, needed in order to view information about several processes being enacted concurrently, as the increased amount of information to be displayed expands the page vertically.

The ability to directly enact processes from within the to-do list service is another important feature that was added. When a stage is ready to be enacted, buttons representing possible enactment choices are displayed next to the stage information in the to-do list. Clicking on one of these enactment buttons triggers the same event as when enacting the corresponding stage via the Pounamu user interface. This feature enhances the tool by providing more flexible and accessible user interaction. Additionally, process enactment access from within the to-do list service is beneficial for possible future incorporation of user-types. For example, low status users might only be allowed to use the to-do list service and not the Pounamu tool itself. Also, because using the to-do list service requires users to supply a valid username and password, their user-types can be determined and enactment facilities provided specifically tailored to the corresponding users.

175

Finally, the to-do list service was altered to comply with the colour scheme employed by IMÅL to show the state and progress of processes in Pounamu. Hence, the tasks displayed in the to-do list are displayed in the appropriate colours corresponding to their states. Stages that are ready to be enacted are shown in red, currently enacted stages are blue, and green is used for completed stages.

Figure 7-18 depicts the improved version of the to-do list service, including functionality to enact processes directly from within the service.

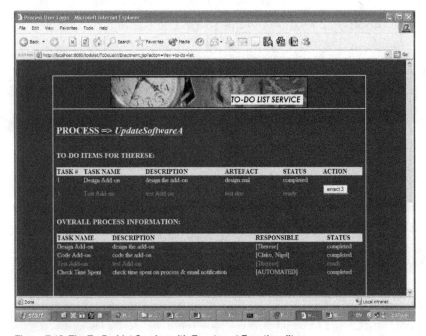

Figure 7-18. The To-Do List Service with Enactment Functionality.

7.8. Summary

The integration of IMÅL with the existing tools Microsoft Infopath and Idiom Decision Suite introduced a new dimension in the environment, namely automation. Both automatic tool activation according to enactment of stages, and automatic completion of stages, are covered by the integration described in this chapter. Although improvements can be made, especially in the way of further distributing the components, the functionality provided by the two new services is generally efficient and useful. Additionally, the enhancement of the to-do list service with some new features has increased the flexibility of the overall system and made its functionality more accessible.

Chapter 8 - Conclusion and Future Work

8.1. Introduction

This chapter concludes the thesis by discussing its major contributions to the software process research area and the software engineering area in general. Suggestions and directions for future work are also presented and elaborated on.

8.2. Essential Contributions of this Thesis

Owing mainly to a novel approach to the development of a process enactment engine, this thesis work makes several contributions to the software process research area. Familiar concepts addressed in this work are also believed to be of value, as another perspective on the concepts and how to deal with them is provided. The major contributions of the development of the IMÅL prototype system are listed below and discussed in the following sections.

- The Pounamu Process Modelling Language.
- The Pounamu Process Modelling Tool.
- A service-oriented process enactment engine.
- A Web-based to-do list service.
- An evaluation of the IMÅL prototype.
- Seamless integration with third-party tools.

179

8.2.1. The Pounamu Process Modelling Language

Developed using the Pounamu meta-CASE tool, this visual process modelling language supports the basic process elements stage, artefact, tool, role, actor and flow. Although simple, the language allows for adequate definition of processes for enactment purposes. Owing to the language being visual, intuitive and easily understandable, it is seen as a valuable addition to process modelling languages.

8.2.2. The Pounamu Process Modelling Tool

This visual process modelling tool was also developed using the Pounamu meta-CASE tool. The tool allows users to model any process by adding stages, role-actor associations and flows, to one or multiple model project views. It is easy to learn, provides intuitive modelling functionality, and stores the process definition in XML-format. The last feature is an advantage over other approaches, as several efficient XML tools are available that provides for simplified interpretation of the process model.

8.2.3. A Service-Oriented Process Enactment Engine

The process enactment engine is the component to which most time and effort was dedicated, and it is consequently seen as the major contribution of the work undertaken. The engine accepts a process model in XML format, parses it to create its own internal process representation, and is then ready to handle enactment events triggered by users. Starting, pausing, resuming and finishing work on stages are the enactment events currently enabled, and the user interaction for enactment is provided by Pounamu and the to-do list service.

The most important novelty of this work is the service-oriented approach to the development of the process enactment engine. It is built as a collection of smaller Web Services, each with their own responsibility in the enactment processing phase. This makes for natural and easy distribution of the system over numerous different locations, while preserving a high level of consistency. Additionally, the engine employs several smaller processing instances assigned to process different types of enactment events,

180

rather than one instance to process all events. An advantage of this approach over the 'one fits all' attitude is that the best of several approaches, such as rule-based or Petri net based, can be utilised to provide one ideal approach. Hence, IMÅL provides a multi-paradigm based process engine. The fact that if one instance fails the others are unaffected serves as another advantage of employing multiple processing instances and several disparate services, as it reduces the impact of a server crash.

Another important feature of the process engine is that it can provide enactment support for processes modelled using any modelling language, as long as the process model is converted to the XML-format the process engine requires. In other words, the engine can be plugged into any process modelling tool, as long as an intermediate translation mechanism is introduced.

8.2.4. A Web-Based To-Do List Service

IMÅL's Web-based to-do list component allows certified users to view individual to-do lists as well as overall process information for enacted processes. The latest version also enables users to enact processes directly from within the service. As it provides users with easy access to information about both their assigned tasks and overall process state, the to-do list service comprises an important contribution of this work. Basically, users can access this information from any computer that has a Web browser and a connection to the Internet. This also eases the coordination of work distributed between users in physically different locations. Additionally, the enactment functionality provided by the to-do list service is a very powerful feature that enhances the flexibility and accessibility of the system considerably.

8.2.5. An Evaluation of the IMÅL Prototype

A Cognitive Dimensions evaluation and a survey employing real users were carried out on the IMÅL system. Both provided useful information and feedback, and serve as a basis for future development and recommendations. As the approach has not frequently been used to evaluate software process technology, and seeing that the Cognitive Dimensions evaluation undertaken shows that useful information can be acquired from such an analysis, this evaluation is seen as a valuable contribution to the research area. This evaluation approach could beneficially be employed by developers of software process technology prior to investing an extensive amount of time and money in user testing. The evaluation survey that was carried out provided useful feedback on the original IMÅL approach, and is therefore also seen as a worthwhile contribution.

8.2.6. Seamless Integration with Third-Party Tools

To address some of the weaknesses identified in the evaluation, some existing external systems were integrated into the IMÅL environment in a seamless fashion. Microsoft Infopath was integrated via a tool service that automatically initialises external tools and documents, whereas Idiom Decision Suite was used to build an automation service that provides automatic processing of process stages. This tool integration serves as an essential contribution, as it shows that the system is highly susceptible to integration of external tools and components, and that such integration can be smoothly achieved.

8.3. Suggestions for Future Work

Several academic prototypes for software process support have been developed, but only some commercial products are available. Considering software companies' and other businesses' increased preoccupation with improving their development processes, this suggests that there is huge potential for further advancement within the software process area. Although the work done in this project is seen as a good start, a considerable amount of work is required in order to convert the current IMÅL prototype into an optimal process support system. Several directions further development could take are listed below, and explained in more detail in the subsequent sections.

- Investigate the applicability of the service-oriented approach.
- Implement additional processing instances.
- Add support for evolution of processes and process models.
- Plug in support for cooperative activities.
- Incorporate different user-types and time concepts.
- Improve support for concurrent enactment.

- Enhance the process modelling functionality.
- Explore the pluggability of the process engine.
- Plug in alternative means for user interaction.

8.3.1. Investigate the Applicability of the Service-Oriented Approach

The usefulness of the service-oriented approach employed in IMÅL's process engine should be examined more closely by integrating other useful Web Services, and possibly also the Universal Description, Discovery and Integration technology which is a means to quickly and dynamically discover and invoke available Web Services on the Internet [64]. Additionally, the Web Services that make up the process engine should be deployed and tested on a highly distributed network, in order to assess the stability and vulnerability of such a distributed and service-oriented architecture. Doing so would also give an indication of the resources required to deploy and use the system, and identify alternative deployment approaches for companies and teams of different sizes.

8.3.2. Implement Additional Processing Instances

Processing simple flow and automatic stages are the responsibilities of the simple flow and automation services respectively. The simple flow service employs an event-based flow strategy, and is only able to make simple flow decisions, usually involving only one variable. Implementing additional processing instances to deal with more complex flow, like decision-making involving several variables and impacting factors, would make the system much more powerful. Additionally, it would become clearer how useful the approach of combining multiple processing instances is. Such a clarification could be of significant importance for the software process research area, and possibly lead to a change towards process engines employing multiple processing instances.

An example of adding another processing instance to IMÅL is to implement and integrate a rule-based complex flow service. Owing to Idiom Decision Suite being rule-based, and because positive experiences have already been gained from using it in the implementation of the automation service, Idiom could beneficially be utilised in such a complex flow context. Figure 8-1 shows a proposed design of such a complex flow service. Basically, it is a complex version of the simple flow service, which utilises Idiom's rule-based approach to make flow decisions. Interface ComplexFlowIF and its implementation class ComplexFlowImpl are the central components of the service, which has its own representation of the process definition created by its local class XMLParser. As in the simple flow service, the classes StateDBClient and GUIandPMRefresherClient are responsible for connecting to and communicating with the state database service and refresher service respectively. Finally, class LocalDecisionServer uses classes generated from a pre-defined specification to make a flow decision based on several variables and other factors involved.

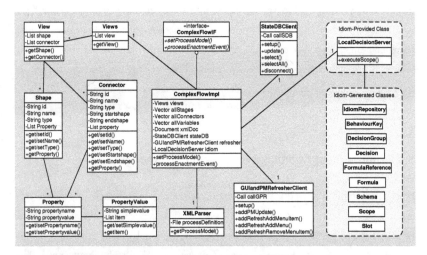

Figure 8-1. Proposed Design of a Complex Flow Service that Utilises Idiom.

Furthermore, additional processing instances with different underlying paradigms, such as extended flow and Petri net based, should also be investigated. When embedding numerous processing instances, it might also be beneficial to examine the deployment of the process engine's internal representation of the process model in a separate service to which all the processing instances have access. This would reduce the duplication of the definition in the engine and make it easier to handle any updates made to the process model, as changes would need to be reflected only in the one process representation service, rather than in all the different processing instances individually.

8.3.3. Add Support for Evolution of Processes and Process Models

A considerable enhancement of the IMÅL environment would be to add support for evolution of processes and process models. In other words, the system should be improved in a way that enables it to deal with changes in processes and process models by reflecting these changes in the system. The reflection of changes should take place without hindering the ongoing enactment of processes, but affect the enactment options according to the changes that are made.

```
/** PMEditedEventListener.java: Interface for listening to process modelling events */
public interface PMEditedEventListener {
  public void receiveModellingEvent(ActionEvent event, DefaultMutableTreeNode node);
}

/** ProcessChange.java: responsible for handling process modelling events */
public class ProcessChange implements PMEditedEventListener {
  PounamuModelProject modelProject;

  /** constructor */
  public ProcessChange(PounamuModelProject modelProject) {
    this.modelProject = modelProject;
    subscribeToPMEditedEvents();
  }

  /** subscribe to listen to process modelling events */
  public void subscribeToPMEditedEvents() {
    modelProject.addPMEditedEventListener(this);
  }

  /** define what happens when a process modelling event is received */
  public void receiveModellingEvent(ActionEvent event, DefaultMutableTreeNode node) {
    // 1. Gain access to and update process representation in engine
    // 2. Apply changes to process model in modelling view
  }
}
```

Figure 8-2. Interface PMEditedEventListener and Class ProcessChange for Process Evolution.

Crucial in supporting process evolution is functionality that allows the user to dynamically edit a process model while it is being enacted. A way this could be achieved is by adding an interface, PMEditedEventListener, which declares a *whenEditedEventFired* method. Another class, ProcessChange, should then implement this interface in order to be notified about events that are fired every time a process model is edited, and define in its *whenEditedEventFired* method how to respond to such events. The method should gain access to, and update, the process engine's internal representation of the process, before

185

applying the changes to the process model in the modelling view. This way, further enactment support is provided on the basis of the altered process definition. Figure 8-2 shows an example of how the PMEditedEventListener interface and class ProcessChange could be implemented.

8.3.4. Plug in Support for Cooperative Activities

Owing to the increased importance of aiding cooperation in software systems [77], support for cooperative activities such as collaborative modelling and enactment of processes should be incorporated in IMÅL to enhance the overall system.

In order to enable collaborative process modelling and enactment, concurrency must be supported, and two issues crucial to allowing concurrency in a system are concurrency control and locking. Concurrency is when two users try to update the same object simultaneously, and concurrency control refers to controlling such a situation. Locking is a way of providing this control, and in the Java language this is accomplished by adding the keyword *synchronized* to methods that need locks [85]. Every object that has synchronised code is then associated with a lock by the Java platform [85].

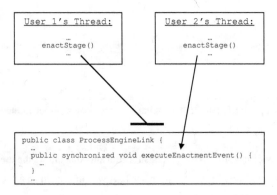

Figure 8-3. Locking of Process Model Objects to Control Concurrent Updates.

Figure 8-3 shows a diagram illustrating how the access to process model objects can be controlled during collaborative modelling and enactment. A user-thread that wants to update an object has to gain access by acquiring the lock before being able to perform the

186

update. If one user-thread is in possession of the lock, another user-thread that wants to update the same object has to wait until the first user-thread releases the lock, either by exiting the method, or by calling the *wait* method.

Finally, adding support for other cooperative activities, such as instant messaging and chatting, would also contribute to enhance the IMÅL environment. Such components could be added onto the system in a similar fashion to the component-based plugging in of the process engine.

8.3.5. Incorporate Different User-Types and Time Concepts

In order for the system to optimally support different types of work environments, the notion of user-types needs to be carefully incorporated into the system. For example, the system should enable user-type 'manager' to be given overall administrative rights, whereas user-type 'employee' only has rights limited to the areas of the process for which he/she is responsible.

The time dimension could also beneficially be integrated into the IMÅL environment, to enhance the overall functionality and applicability of the system. As explained in Section 6.3.4, the incorporation of time concepts, in terms of tracking the time spent on processes and parts of them, could be done using Pounamu event handlers.

8.3.6. Improve Support for Concurrent Enactment

By enabling users to plug in a process engine for each process model, IMÅL allows for multiple processes to be enacted simultaneously. However, there is a need to also support multiple people working concurrently on the same task. The latter can be done in association with the implementation of user-types. For example, if two users are assigned to a stage and one of them has enacted it, the other user should still be presented with a possibility to enact the same stage. This would have to be implemented by adjusting the refresher service, so that different enactment options are presented to different users depending on their user-type. Concurrency issues, as explained earlier, need also be taken into consideration when implementing this.

8.3.7. Enhance the Process Modelling Functionality

To make the process modelling functionality easier to use and less error-prone, it is necessary to provide a mechanism that automatically generates unique identifiers for stages modelled by users. A way this could be achieved is by implementing a counter that keeps track of how many stages have been modelled, and assigns identifiers corresponding to the order of creation. Furthermore, this mechanism should be extended to also provide a way for users to open, within a parent stage, sub-views that are automatically named after their parent. This would make life easier for the user, and at the same time make the process engine's interpretation of the process model simpler.

Feedback from the evaluation survey also highlighted the need to add support for modelling of multiple artefacts and tools for each stage. This can be achieved simply by changing the definition of the process modelling tool in Pounamu, but the process enactment engine will need to be updated correspondingly.

8.3.8. Explore the Pluggability of the Process Engine

Another direction further work could take is to try plugging the process engine into secondary process modelling tools. As mentioned earlier, an intermediate step that translates the original process model into the format that the process engine requires would have to be added, in order for the process engine to work in such a situation. Testing the process engine in such a way would give an idea of how flexible the process engine is in terms of pluggability, and whether it can beneficially be used independently of the process modelling tool.

8.3.9. Plug in Alternative Means for User Interaction

Developing and utilising additional means for process modelling and user interaction with the system would also enhance the IMÅL tool. As opposed to the thick-client approach employed so far, a useful extension to the system would be to deploy a thin-client user interface for Pounamu. This would mean that users would not need a copy of the Pounamu software installed on their computers to be able to model processes and run the

188

system. Instead, users would only need a Web browser in which to run the thin-client version of Pounamu, while the main Pounamu is hosted by one or several central servers. An important consideration when implementing such a thin-client interface for using the tool is security. Adequate security measures are essential, in order to avoid uncertified users accessing the Pounamu interface and performing unwanted actions. The work on a thin-client version of Pounamu is an ongoing thesis project of another student in the Computer Science Department at the University of Auckland, Penny Cao.

8.4. Summary

Since Manny Lehman initiated his study of the IBM programming process in 1969 [52], more and more attention has been awarded to the software process research area. This thesis has focused on the support of ongoing software processes by describing the development of the IMÅL prototype process management and support system. The most essential contribution of this work is the novel, service-oriented process enactment engine employing multiple processing instances. Other important contributions are the to-do list service and the tool integration services, as well as an evaluation of the prototype built. Finally, the work has also provided the software process area with a new visual process modelling language and tool.

Several openings for further development and research arise as a result of the work undertaken in this thesis project. Further work on the process engine and its architecture can advantageously be carried out, for example, by more closely investigating its service-oriented approach, adding additional processing instances and investigating the engine's pluggability. Also, the support for cooperative activities, process evolution and concurrent enactment in IMÅL are all aspects of importance that should be followed up on. Finally, the process modelling functionality can beneficially be improved, and alternative means for user interaction with the system should be explored, such as for example a thin-client user interface for Pounamu.

Appendix A – Tool Tutorial

TOOL TUTORIAL

This tutorial shows you how to use the IMÅL process support and management system. The user interaction with the system is provided by the Pounamu software tool. In order to enhance understanding of how to use the system, the first section briefly introduces the Pounamu tool and its concepts as well as the Pounamu Process Modelling Tool (PPMT) built for this project. User tasks are presented in the succeeding section.

A Brief Introduction to the Pounamu Software Tool and PPMT

Pounamu is a visual software engineering tool that allows users to create software engineering tools as well as model software engineering projects. In this project, Pounamu's tool creator has been used to create the PPMT, and Pounamu's project modeller is used to model and enact processes using PPMT.

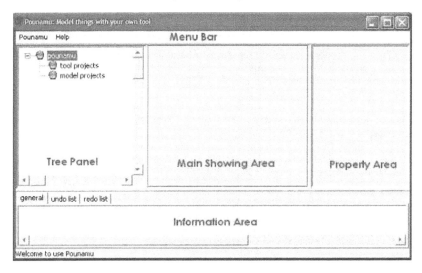

A screen dump of the Pounamu user interface on start-up is shown above, with the main components marked in red. The main components are:

1. **The main menu bar** – contains access to top-level functionality.
2. **The tree panel** – manages opened projects and their entities.
3. **The main showing area** – displays selected projects and entities.
4. **The property area** – shows the properties of selected entities.
5. **The information area** – displays modelling-related information.

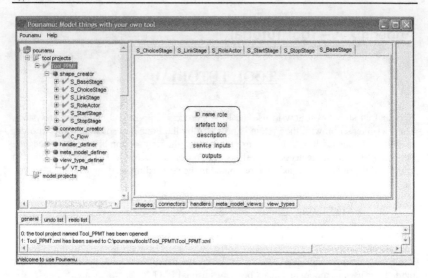

Depicted above, PPMT uses an underlying process modelling language consisting of seven elements (six Pounamu shapes and one Pounamu connector) in order to enable process modelling. StartStage, StopStage, BaseStage and RoleActor are the four most important shapes. StartStage represents the start of a process, StopStage represents the conclusion of a process, BaseStage represents work to be done, and RoleActor associates human users with roles. The connector, Flow, corresponds to flow between stages. PPMT has one defined view-type, namely VT_PM, which enables users to use all the defined elements to model processes.

Element BaseStage is the most complex of the elements in the process modelling language, as it is the stage that contains work to be done, and therefore holds information about the work. The design of the element, including its shape and attributes, is shown in the main showing area in the figure above. The following lists and describes the attributes of element BaseStage:

- ID = the stage identifier (unique)
- name = the name of the stage
- description = a description of the work involved
- role = the role responsible for performing the work
- artefact = the artefact involved
- tool = the tool used
- service = the service responsible for processing this stage
- inputs = a list of possible inputs leading to the enactment of the stage
- outputs = a list of possible outputs resulting from completion of the stage

The Tasks

The tutorial is split up into five main tasks as follows: open a new model project, model a software process, plug in the process engine, enact the process model, and view to-do list.

Each task has sub-tasks explained in the respective sections, and concludes with the result of performing these tasks. The process that is modelled and enacted in this tutorial is a simplified version of a process in which the aim is to *add functionality to an existing software program*. Please work through all the tasks sequentially before completing the questionnaire.

For simplicity and anonymity reasons, please use 'Myself' to refer to you when modelling processes. Further, your username for the to-do list service is 'Myself' and your password is 'gold'.

1. Open a new model project using PPMT:

Start a new model project dialogue by selecting and right-clicking 'model projects' in the tree panel, and choosing 'new model project' from the drop-down menu.

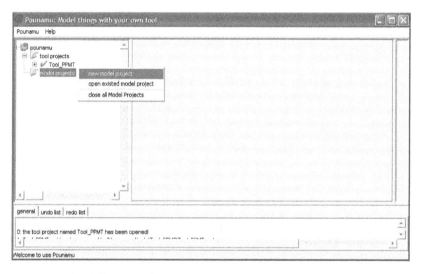

Input 'UpdateSoftware' as the name for the new model project and 'A simple process that adds functionality to an existing software program' as the description. Further, choose 'Tool_PPMT' as the tool to be used for the new model project, and click OK.

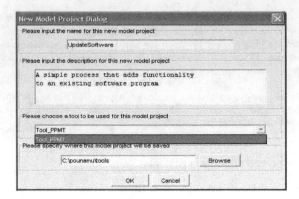

RESULT: A model project called 'Model_UpdateSoftware', in which a process can be modelled, has been opened.

2. Model a process:

Click on the '+' to the left of the 'Model_UpdateSoftware' node in the tree panel to expand the model project. There is one view-type associated with this model project, namely 'VT_PM'.

Open a new view of view-type 'VT_PM' by selecting and right-clicking the 'VT_PM' node, and choosing 'open a new VT_PM view' from the drop-down menu.

Input 'AddOnProcess' as the name for the view in the new model view name dialogue. The 'VT_PM_AddOnProcess' view is created and represented by a node in the tree panel.

Add stages to the view:

Add a start-stage to the view 'VT_PM_AddOnProcess' by selecting and right-clicking the 'VT_PM_AddOnProcess' node, and choosing 'add an object of ET_StartStage' from the drop-down menu.

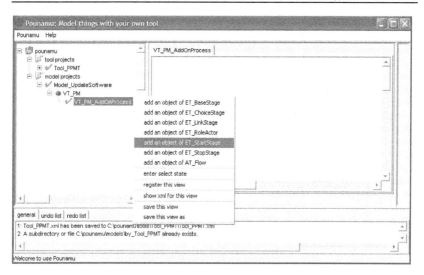

Input a name for the start-stage in the object dialogue and click OK.

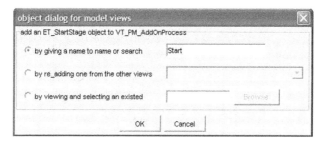

Click in the top left hand corner of the main canvas to add the start-stage in that location. The element is added as a new node in the tree panel, under the 'VT_PM_AddOnProcess' view. (You can add the stage at any location in the canvas. This tutorial adds it in the top left hand corner, in order to make room for the other elements that are to be added to the view.)

Save the stage by selecting 'save this entity' from its drop-down menu in the tree panel.

Using the same approach as in steps 0-0, add and save a stop-stage ('ET_StopStage') to the bottom right hand corner of the canvas.

It is now time to add the work-stages of the process to the view. Again, use the same approach as in steps 0-0 to add and save a work-stage ('ET_BaseStage') to the canvas.

In the right hand property panel, add values to the stage properties, and click OK at the bottom of the property panel to save the information. Add the following values:

195

ID = 1
artefact = design.vsd
description = design the add-on
inputs = start
name = Design Add-On
outputs = design completed
role = Designer
service = simpleflow
tool = MS Visio

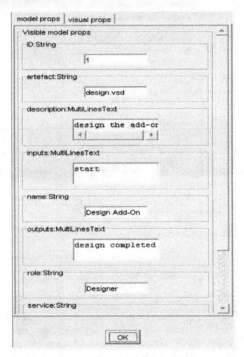

Add and save two more work-stages with property values as specified in the table below. (You can ignore the first row of the table, as it contains the first work-stage which you have already added.)

id	artefact	description	inputs	name	outputs	role	service	tool
1	design.vsd	design the add-on	-start	Design Add-On	-design completed	Designer	simpleflow	visio
2	code.java	code the add-on	-design completed -test failed	Code Add-On	-code completed	Coder	simpleflow	jcreator
3	test.doc	test the add-on	-code completed	Test Add-On	-test succeeded -test failed	Tester	simpleflow	word

NOTE: When adding the values for the *inputs* and *outputs* fields, make sure each value is entered on a new line in the MultiLinesText area in the property panel (each value is marked with a '-' in the table above).

The figure below shows approximately what your view should look like after completing all the steps so far. Make sure that you have saved all the stages (indicated with pink) you have added to the view.

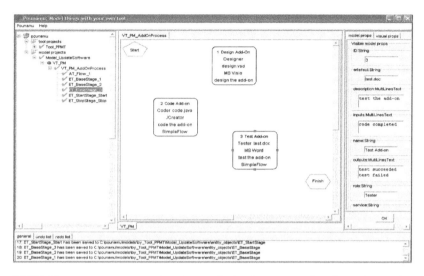

NOTE: If you want to rearrange the elements in the view, you need to choose 'enter select state' from the drop-down menu of node 'VT_PM_AddOnProcess'. This allows you to select and move the elements around the canvas.

Add flows to the view:

Add a flow to the view by selecting and right-clicking the 'VT_PM_AddOnProcess' node, and choosing 'add an object of AT_Flow' from the drop-down menu.

Input a name for the flow in the object dialogue, and click OK.

Add the flow to start from the start-stage, and end in stage 1, by clicking on one of the handlers on the start-stage, dragging the mouse (hold the mouse button down while dragging), and releasing it on one of the handlers on stage 1.

Save the flow by selecting 'save this association' from its drop-down menu in the tree panel.

Using the same approach as in steps 0-0, add four more connectors as follows:

-   starting from stage 1 and ending in stage 2
-   starting from stage 2 and ending in stage 3
-   starting from stage 3 and ending in stage 2
-   starting from stage 3 and ending in the stop-stage

The figure below shows approximately what your view should look like after adding all the flows. Make sure that you have saved all the flows (indicated with pink) you have added to the view.

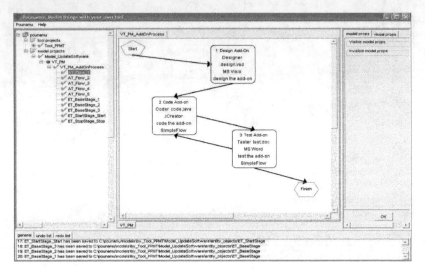

Add a role-actor associations to the view:

Add a role-actor association to the view by choosing 'add an object of ET_RoleActor' from the view's drop-down menu. (Alternatively, you can open a new view, and add the role-actor associations to the new view. For simplicity, this tutorial adds them to the same view as the rest of the process model.)

Input a name for the role-actor association in the object dialogue, and click OK.

Click somewhere in the main canvas to add the role-actor association in that location.

In the right hand property panel, add values to the role-actor association properties, and click OK at the bottom of the property panel to save the information. Add the following values:

actors = Myself
role = Designer

Save the role-actor association by selecting 'save this entity' from its drop-down menu in the tree panel.

Following the same approach as in steps 0-0, add two more role-actor associations to the view with property values as specified in the table below. (You can ignore the first row of the table, as it contains the first role-actor association which you have already added.)

role	actors
Designer	-Myself

Coder	-Bob
	-Amy
Tester	-Myself

NOTE: When adding the values for the *actors* field, make sure each value is entered on a new line in the MultiLinesText area in the property panel (each value is marked with a '-' in the table above).

The figure below shows approximately what your view should look like after adding all the role-actor associations. Make sure that you have saved all the role-actor associations (indicated with pink) you have added to the view.

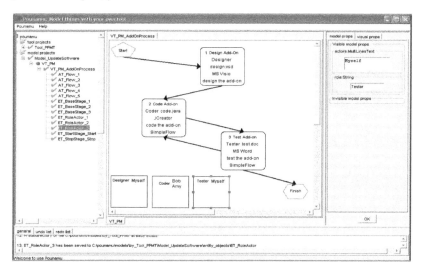

Save the view by choosing 'save this view' from the drop-down menu of node 'VT_PM_AddOnProcess'.

Save the model project by choosing 'save this model' from the drop-down menu of node 'Model_UpdateSoftware'.

RESULT: The process model is complete, and the process engine can be plugged in to enable enactment.

3. Plug in process engine:

From the drop-down menu of node 'Model_UpdateSoftware', choose 'plug in process engine for this model project'.

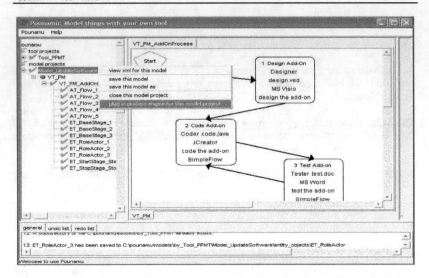

RESULT: The process engine is plugged in and initialised, and the process model is ready for enactment.

4. Enact the process model:

From the drop-down menu of node 'VT_PM_AddOnProcess', choose 'enter select state'.

Right-click on the start-stage in the main view, and choose 'start process' from the drop-down menu. The user interface and process model is now updated according to the process definition, and guides you through the process. The stage colours correspond to the state of stages as follows:

- RED = ready
- BLUE = enacted
- GREEN = completed

Enact stage 1 by right-clicking it, and choosing 'enact stage' from the drop-down menu.

Pause stage 1 by right-clicking it, and choosing 'pause stage' from the drop-down menu.

Resume stage 1 by right-clicking it, and choosing 'resume stage' from the drop-down menu.

Finish stage 1 in 'design completed' mode by right-clicking it, and choosing 'design completed' from the submenu of 'finish stage' on the drop-down menu.

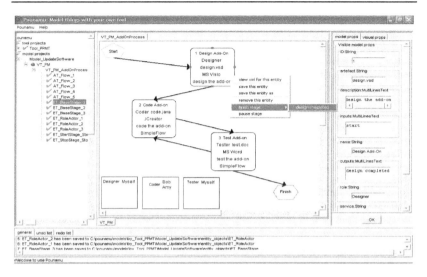

RESULT: The process model is enacted, and certified users can make use of the to-do list service in order to view individual to-do lists for the process.

5. View to-do list:

In a Web browser, open the URL 'http://localhost:8080/todolist'.

Enter your username ('Myself') and password ('gold') into the username and password text fields and click the 'Login' button.

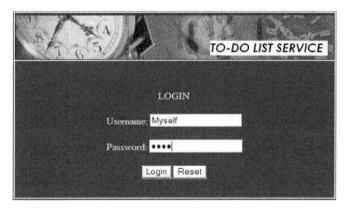

Click the 'View to-do list' button on the next page.

Go back to Pounamu and enact stage 2 by right-clicking it and choosing 'enact stage' from the drop-down menu.

Go back to the Web browser to see that the to-do list has been updated according to the enactment in step 0.

RESULT: A to-do list showing the work you are responsible for in this process is displayed, and automatically updated every 20 seconds.

6. Finally, complete the process by enacting and finishing the remaining stages.

If you wish to further explore the tool by modelling your own processes and enacting them, please feel free to do so. You can open a new model project using PPMT, model a new process, and plug in a process engine for the new process model in order to enable enactment. In other words, the tool supports the running of multiple processes simultaneously, by plugging in a new process engine for each process model.

You can remove the process engine from any model project by selecting and right-clicking the model project node, and choosing 'remove process engine from this model project' from the drop-down menu. Removing the process engine restores the initial state of the model project (as before the process engine was plugged in). Thus, the menu items that enable enactment are removed, and the original colours of the process model are restored.

A way of resetting the process without removing the process engine from the model project is also provided. Choosing 'clear process' from the drop-down menu of the start-stage in the process restores the process model to the state it was in before enactment. However, enactment menu items are not removed, and you can start the process again from the start-stage.

Appendix B – Questionnaire

QUESTIONNAIRE

The purpose of this questionnaire is to collect feedback on the IMÅL prototype process support system that has been built. The feedback will be used when analysing and evaluating the system and its applicability, and as a basis for suggesting future improvement of the tool.

All the information you provide will remain anonymous, and will not be used for any other purpose than that described above.

When answering the questions, please keep in mind that the tool is only a prototype and can therefore not be compared with commercial products. Try to focus on the functionality the tool provides rather than aspects not implemented in this prototype.

Please answer all the questions below. Answer the ranking and the yes/no questions by circling the number that corresponds to your answer. For the remaining questions, write your answers in the spaces provided.

Process Modelling:

How easy was it to use the process modelling tool?

Very easy		Average		Very difficult
1	2	3	4	5

How intuitive was the process modelling language (i.e. the shapes and connectors available for modelling processes)?

Very intuitive		Average		Not at all intuitive
1	2	3	4	5

Do you have any other comments on the process modelling functionality provided by the tool?

_____

_____

_____

_____

_____

_____

_____

_____

Process Enactment:

How easy was it to enact the process model when the process engine was plugged in?

Very easy		Average		Very difficult
1	2	3	4	5

How satisfactory was the guidance on the status of the process and which enactment-options you had available at any time (via stage colour and menu item changes)?

Very satisfactory		Average		Not at all satisfactory
1	2	3	4	5

How useful do you think it is that stage colour changes according to process enactment?

Very useful		Average		Not at all useful
1	2	3	4	5

Do you have any other comments on the process enactment functionality provided by the tool?

_____

_____

_____

_____

_____

_____

_____

_____

View To-Do List:

How well did you think the to-do list service provided information about the progress of the process?

Very well		Average		Not at all well
1	2	3	4	5

How relevant do you think it is to provide the users of the tool with such a to-do list service?

Very relevant		Average		Very irrelevant
1	2	3	4	5

Do you have any other comments on the to-do list service?

_____

_____

_____

_____

_____

_____

_____

_____

Miscellaneous:

In general, do you think the tool is useful in supporting processes?

Yes   No

Provided the prototype is evolved into a fully working and robust system, would you consider using it in your office/business?

Yes   No

Do you have any suggestions for improvement of the system?

_____

_____

_____

_____

_____

_____

_____

_____

Do you have any other comments about the system (positive/negative)?

_____

_____

_____

_____

_____

_____

_____

_____

List of References

1. S.W. Ambler. "An Introduction to Process Patterns." White Paper, Ambysoft Inc., June 27 1998. 18 p.
   http://www.ambysoft.com/processPatterns.pdf

2. S.W. Ambler. "Process Patterns: Building Large-Scale Systems Using Object Technology." Cambridge University Press/SIGS Books, New York - USA, 1998. ISBN: 0-521-64568-9. 549 p.

3. S.W. Ambler. "More Process Patterns: Delivering Large-Scale Systems Using Object Technology." Cambridge University Press/SIGS Books, New York - USA, 1999. ISBN: 0-521-65262-6. 416 p.

4. Apache Software Foundation. "The Apache Jakarta Project." 2003.
   http://jakarta.apache.org/tomcat/

5. S.C. Bandinelli, E. Di Nitto, and A. Fuggetta. "Supporting Cooperation in the SPADE-1 Environment." *IEEE Transactions of Software Engineering*, Vol. **22**(No. 12), December 1996. p. 841-865.

6. L. Bass, P. Clements, and R. Kazman. "Software Architecture in Practice." Addison-Wesley, Boston, Massachussetts - USA, 1997. ISBN: 0-201-19930-0. 480 p.

7. K. Beck. "Embracing Change with Extreme Programming." *IEEE Computer*, Vol. **32**(No. 10), October 1999. p. 70-77.

8. K. Beck. "Extreme Programming Explained: Embrace Change." Addison-Wesley, Massachusetts - USA, 1999. ISBN: 0-201-61641-6. 224 p.

9. I.Z. Ben-Shaul and G.E. Kaiser. "A Paradigm for Decentralized Process Modeling." Kluwer Academic Publishers, Boston - USA, 1995. ISBN: 0-7923-9631-6. 320 p.

10. I.Z. Ben-Shaul and G.E. Kaiser. "Integrating Groupware Activities into Workflow Management Systems." *In Proc. 7th Israeli Conference on Computer-Based Systems and Software Engineering*, Herzliya - Israel, June 10-12 1996, IEEE CS Press. p. 140-149.

11. A.F. Blackwell, C. Britton, A. Cox, T.R.G. Green, C. Gurr, G. Kadoda, M.S. Kutar, M. Loomes, M. Nehaniv, M. Petre, C. Roast, C. Roes, A. Wong, and R.M. Young. "Cognitive Dimensions of Notations: Design Tools for Cognitive

Technology." *In Proc. Cognitive Technology 2001*, Warwick - UK, August 6-9 2001, LNCS 2117, Springer-Verlag. p. 325-341.

12. B.W. Boehm. "Software Engineering." *IEEE Transactions on Computers*, Vol. **25**(No. 12), December 1976. p. 1226-1241.

13. B.W. Boehm. "Software Engineering Economics." Prentice Hall, Englewood Cliffs, New Jersey - USA, 1981. ISBN: 0-138-22122-7. 767 p.

14. B.W. Boehm. "A Spiral Model of Software Development and Enhancement." *IEEE Computer*, Vol. **21**(No. 5), May 1988. p. 61-72.

15. D.P. Bogia and S.M. Kaplan. "Flexibility and Control for Dynamic Workflows in the wOrlds Environment." *In Proc. Conference on Organizational Computing Systems*, Milpitas, California - USA, August 13-16 1995, ACM Press. p. 148-161.

16. C. Bunse, M. Pelayo, and J. Zettel. "Out of the Dark: Adaptable Process Models for XP." *In Proc. XP Conference*, Genova - Italy, May 26-30 2002. p. 109-112.

17. K.M. Carley and Z. Lin. "Organizational Designs Suited to High Performance under Stress." *IEEE Transactions on Systems, Man and Cybernetics*, Vol. **25**(No. 2), February 1995. p. 221-230.

18. P.P. Chen. "The Entity-Relationship Model: Towards a Unified View of Data." *ACM Transactions on Database Systems*, Vol. **1**(No. 1), January 1976. p. 9-36.

19. E.F. Codd. "A Relational Model for Large Shared Data Banks." *Communications of the ACM*, Vol. **13**(No. 6), June 1970. p. 377-387.

20. R. Conradi and M.L. Jaccheri (editors). "Process Modelling Languages." in "Software Process: Principles, Methodology, and Technology." ed. J.C. Derniame, B.A. Kaba, and D. Wastell. LNCS 1500, Springer-Verlag, 1999. p. 27-52.

21. R. Conradi, M.L. Jaccheri, C. Mazzi, N.N. Minh, and A. Aarsen. "Design, Use, and Implementation of SPELL, a Language for Software Process Modeling and Evolution." *In Proc. European Workshop on Software Process Technology (EWSPT)*, Trondheim, Norway, September 7-8 1992, LNCS 635, Springer-Verlag. p. 167-177.

22. R. Conradi and C. Liu. "Process Modelling Languages: One or Many?" *In Proc. 4th European Workshop on Software Process Technology*, Noordwijkerhout - The Netherlands, April 3-5 1995, Springer-Verlag. p. 98-118.

23. J.W. Cooper. "The Design Patterns Java Companion." Addison-Wesley, 1998. 218 p.

24. J.O. Coplien. "A Generative Development-Process Pattern Language." in "Pattern Languages of Program Design." ed. J.O. Coplien and D.C. Schmidt, Addison-Wesley, Massachusetts - USA, 1995. p. 184-237.

25. J.O. Coplien and D.C. Schmidt eds. "Pattern Languages of Program Design." Addison-Wesley, Massachusetts - USA, 1995. ISBN: 0-201-60734-4. 562 p.

26. B. Curtis, M. Kellner, and J.W. Over. "Process Modelling." *Communications of the ACM*, Vol. **35**(No. 9), September 1992. p. 75-90.

27. M. Dowson and C. Fernström. "Towards Requirements for Enactment Mechanisms." *In Proc. 3rd European Workshop on Software Process Technology*, Villard de Lans - France, February 7-9 1994, LNCS 722, Springer-Verlag. p. 90-106.

28. C.A. Ellis, S.J. Gibbs, and G.L. Rein. "Groupware: Some Issues and Experiences." *Communications of the ACM*, Vol. **34**(No. 1), January 1991. p. 38-58.

29. W. Emmerich (editor), A. Finkelstein, A. Fuggetta, C. Montangero, and J.C. Derniame. "Software Process - Standards, Assessments and Improvement." in "Software Process: Principles, Methodology, and Technology." ed. J.C. Derniame, B.A. Kaba, and D. Wastell. LNCS 1500, Springer-Verlag, 1999. p. 15-25.

30. European Space Agency. "ESA Software Engineering Standards." ESA PSS-05-0, Issue 2, European Space Agency, Paris - France, February 1991.

31. C. Fernström. "Process Weaver: Adding Process Support to Unix." *In Proc. 2nd International Conference on the Software Process*, Berlin - Germany, February 25-26 1993, IEEE CS Press. p. 12-26.

32. M.S. Fox and M. Gruninger. "Enterprise Modelling." The AI Magazine, Vol. **19**(No. 3), Fall 1998. p. 109-121.

33. J. Galbraith. "Designing Complex Organizations." Addison-Wesley, UK, 1973. ISBN: 0-201-02559-0. 150 p.

34. E. Gamma, R. Helm, R. Johnson, and J. Vlissides. "Design Patterns: Abstraction and Reuse of Object-Oriented Design." *In Proc. 7th European Conference on Object-Oriented Programming*, Kaiserslautern - Germany, July 26-30 1993, LNCS 707, Springer-Verlag. p. 406-431.

35. E. Gamma, R. Helm, R. Johnson, and J. Vlissides. "Design Patterns: Elements of Reusable Object-Oriented Software." Addison-Wesley, Reading, Massachusetts - USA, 1995. ISBN: 0-201-63361-2. 416 p.

36. J. Gray. "Software Engineering Tools." *In Proc. 33rd Hawaii International Conference on System Sciences*, Maui - Hawaii, January 4-7 2000, 8, IEEE CS Press.

37. T.R.G. Green. "Cognitive Dimensions of Notations." in "People and Computers V." ed. A. Sutcliffe and L. Macaulay, Cambridge University Press, Cambridge - UK, 1989. p. 443-460.

38. T.R.G. Green. "An Introduction to the Cognitive Dimensions Framework." 1996. http://www.thomas-green.ndtilda.co.uk/workStuff/Papers/introCogDims/index.html

39. J.C. Grundy, M.D. Apperley, J.G. Hosking, and W.B. Mugridge. "A Decentralized Architecture for Software Process Modeling and Enactment." *IEEE Internet Computing*, Vol. **2**(No. 5), September-October 1998. p. 53-62.

40. J.C. Grundy and J.G. Hosking. "Serendipity: Integrated Environment Support for Process Modelling, Enactment and Work Coordination." *Automated Software Engineering*, Vol. **5**(No. 1), January 1998. p. 27-60.

41. J.C. Grundy and J.G. Hosking. "Engineering Plug-In Software Components to Support Collaborative Work." *Software - Practice and Experience*, Vol. **32**(No. 10), August 2002. p. 983-1013.

42. J.C. Grundy, J.G. Hosking, and W.B. Mugridge. "Support for End-User Specification of Work Coordination in Workflow Systems." *In Proc. 2nd International Workshop on End-User Development*, Barcelona - Spain, June 16-17 1997.

43. J.C. Grundy, J.G. Hosking, and W.B. Mugridge. "Coordinating Distributed Software Development Projects with Integrated Process Modelling and Enactment Environments." *In Proc. 7th IEEE Workshop on Enabling Technologies: Infrastructure for Collaborative Enterprises*, Palo Alto, California - USA, June 17-19 1998, IEEE CS Press. p. 39-44.

44. M. Hammer and J. Champy. "Reengineering the Corporation: A Manifesto for Business Revolution." Harper Business Press, New York - USA, 1993. ISBN: 0-88730-687-X. 233 p.

45. T. Helland. "Collaborative Editing in the Pounamu Environment." CompSci780 Project Report, University of Auckland, Auckland - New Zealand, October 25 2002. 54 p.

46. Idiom Software Ltd. "Idiom." 2001.

http://www.idiomsoftware.com/index.asp

47.   Idiom Software Ltd. "Idiom Decision Server v.2.1 - USER MANUAL." Idiom Ltd, October 2002. 162 p.

48.   IEEE Computer Society. "IEEE Standard Glossary of Software Engineering Terminology." IEEE Standard 610.12-1990, The Institute of Electrical and Electronics Engineers, Piscataway, New Jersey - USA, December 1990. 82 p.

49.   M.L. Jaccheri, J.O. Larsen, and R. Conradi. "Software Process Modeling and Evolution in EPOS." *In Proc. 4th International Conference on Software Engineering and Knowledge Engineering*, Capri - Italy, June 17-29 1992, IEEE CS Press. p. 574-581.

50.   D.E. Knuth. "The Art of Computer Programming." Vol. 1: "Fundamental Algorithms". Addison-Wesley, Reading, Massachusetts - USA, 1968. ISBN: 0201896834. 672 p.

51.   P. Kruchten. "The Rational Unified Process: An Introduction (2nd Edition)." Addison-Wesley, 2000. ISBN: 0201707101. 320 p.

52.   M.M. Lehman. "The Programming Process." RC 2722, Report, IBM Research Centre, Yorktown Heights, New York - USA, December 1969. 47 p.

53.   M.M. Lehman and L.A. Belady. "Program Evolution: Processes of Software Change." Academic Press, New York - USA, 1985. ISBN: 0-12-442441-4. 538 p.

54.   J. Lonchamp. "An Assessment Exercise." in "Software Process Modelling and Technology." ed. A. Finkelstein, J. Kramer, and B. Nuseibeh, Research Studies Press, London - UK, 1994. p. 335-356.

55.   R. Malan and D. Bredemeyer. "Defining Non-Functional Requirements." White Paper, Bredemeyer Consulting, March 8 2001. 8 p.

www.bredemeyer.com/pdf_files/NonFunctReq.PDF

56.   F. Maurer, B. Dellen, F. Bendeck, S. Goldmann, H. Holz, B. Kötting, and M. Schaaf. "Merging Project Planning and Web-Enabled Dynamic Workflow Technologies." *IEEE Internet Computing*, Vol. **4**(No. 3), May-June 2000. p. 65-74.

57.   S. McPherson. "JavaServer Pages: A Developer's Perspective." April 2000.

http://developer.java.sun.com/developer/technicalArticles/Programming/jsp/

58.   Microsoft. "Microsoft Access - The Office XP Database Solution." 2003.

http://www.microsoft.com/office/access/default.asp

59.   Microsoft. "Microsoft Office Infopath Product Guide." March 10 2003. 57 p.

60. Microsoft. "Microsoft SQL Server - The Enterprise Relational Database Management and Analysis System." 2003.
http://www.microsoft.com/sql/

61. C. Montangero (editor), J.C. Derniame, B.A. Kaba, and B. Warboys. "The Software Process: Modelling and Technology." in "Software Process: Principles, Methodology, and Technology." ed. J.C. Derniame, B.A. Kaba, and D. Wastell. LNCS 1500, Springer-Verlag, 1999. p. 1-13.

62. P. Naur and B. Randell (editors). "Software Engineering: A Report on a NATO Conference Sponsored by the NATO Science Committee." *In Proc. NATO Conference on Software Engineering*, Garmisch-Partenkirchen - Germany, 7-11 October 1968, NATO Scientific Affairs Division, Brussels, January 1969. 231 p.

63. J.J. Navarro. "Characteristics of a Flexible Software Factory: Organization Design Applied to Software Reuse." *In Proc. 38th IEEE Computer Society International Conference (COMPCON)*, San Francisco, California - USA, February 22-26 1993. p. 265-267.

64. OASIS. "Universal Description, Discovery and Integration of Web Services." 2003.
http://www.uddi.org/

65. Object Management Group. "CORBA." 2003.
http://www.corba.org/

66. F. Oquendo. "Call for Papers - 9th European Workshop on Software Process Technology." 2003.
http://www.cs.kuleuven.ac.be/~dirk/ada-belgium/events/03/030901-ewspt.html

67. Oracle Corporation. "Oracle Database - The World's Most Popular Database." 2003.
http://www.oracle.com/ip/deploy/database/oracle9i

68. L. Osterweil. "Software Processes are Software too." *In Proc. 9th International Conference on Software Engineering*, Monterey, California - USA, March 30 - April 2 1987, IEEE CS Press. p. 2-13.

69. L. Osterweil. "Software Processes are Software too, Revisited: An Invited Talk on the Most Influential Paper of ICSE9." *In Proc. 19th International Conference on Software Engineering*, Boston, Massachusetts - USA, May 17-23 1997, ACM Press. p. 540-548.

70.   B. Peuschel and W. Schäfer. "Concepts and Implementation of a Rule-Based Process Engine." *In Proc. 14th International Conference on Software Engineering*, Melbourne - Australia, May 1992, IEEE CS Press. p. 262-279.

71.   C. Potts (editor). "Proceedings of a Software Process Workshop." *In Proc. 1st International Software Process Workshop*, Egham - UK, February 1984, IEEE CS Press. p. 3-155.

72.   T. Quatrani. "Introduction to the Unified Modeling Language." White Paper, Rational Software, 2001. 20 p.
      http://www.rational.com/media/uml/intro_rdn.pdf

73.   B. Randell and J.N. Buxton (editors). "Software Engineering Techniques: Report of a Conference Sponsored by the NATO Science Committee." *In Proc. NATO Conference on Software Engineering Techniques*, Rome - Italy, 27-31 October 1969, Scientific Affairs Division, Brussels, April 1970. 164 p.

74.   U. Rembold, B.O. Nnaji, and A. Storr. "Computer Integrated Manufacturing and Engineering." Addison-Wesley, Wokingham - UK, 1993. ISBN: 0-201-56541-2. 640 p.

75.   W.W. Royce. "Managing the Development of Large Software Systems." *In Proc. IEEE WESCON*, California - USA, August 1970. p. 1-9.

76.   W. Scacchi. "Process Models in Software Engineering." in "Encyclopedia of Software Engineering." ed. J.J. Marciniak. December, John Wiley & Sons, New York - USA, 2001.

77.   E.M. Schooler. "Conferencing and Collaborative Computing." *Multimedia Systems*, Vol. 4(No. 5), October 1996. p. 210-225.

78.   I. Sommerville. "Software Engineering." 6th ed. Addison-Wesley, 2000. ISBN: 0-201-39815-X. 693 p.

79.   Sun Microsystems. "What is Java Technology?" 2002.
      http://java.sun.com/java2/whatis/

80.   Sun Microsystems. "Developing XML Solutions with JavaServer Pages Technology." 2003.
      http://java.sun.com/products/jsp/html/JSPXML.html

81.   Sun Microsystems. "Java Architecture for XML Binding (JAXB)." 2003.
      http://developer.java.sun.com/developer/technicalArticles/WebServices/jaxb/

82.   Sun Microsystems. "Java Language Overview." 2003.
      http://java.sun.com/docs/overviews/java/java-overview-1.html

83.   Sun Microsystems. "Java Remote Method Invocation (RMI)." 2003.
      http://java.sun.com/products/jdk/rmi/

84.   Sun Microsystems. "Java Servlet Technology." 2003.
      http://java.sun.com/products/servlet/

85.   Sun Microsystems. "Synchronizing Threads." 2003.
      http://java.sun.com/docs/books/tutorial/essential/threads/multithreaded.html

86.   S.M. Sutton. "APPL/A: A Prototype Language for Software-Process Programming." PhD, University of Colorado, Colorado - USA, 1990.

87.   K.D. Swenson. "A Visual Language to Describe Collaborative Work." *In Proc. IEEE Symposium on Visual Languages*, Bergen - Norway, August 24-27 1993, IEEE CS Press. p. 298-303.

88.   K.D. Swenson. "A Business Process Environment Supporting Collaborative Planning." *Journal of Collaborative Computing*, Vol. **1**(No. 1), January 1994. p. 15-34.

89.   Systinet Corporation. "Introduction to Web Services." Cambridge, Massachussetts, USA, 2003. 13 p.

90.   Texcel Systems Inc. "FormBridge - The Fastest Way to Convert Your Forms to E-Forms." 2003.
      http://www.texcel.com

91.   The University of Auckland Human Subjects Ethics Committee. "Guidelines for Applicants." 1999.
      http://www.auckland.ac.nz/docs/research/99%20Guidelines%20Revision.doc

92.   T. Totland and R. Conradi. "A Survey and Classification of Some Research Areas Relevant to Software Process Modeling." *In Proc. 4th European Workshop on Software Process Technology*, Noordwijkerhout - Netherlands, April 3-5 1995, Springer-Verlag. p. 65-70.

93.   D. Van Camp. "The Object-Oriented PatternDigest." 2002.
      http://www.patterndigest.com/

94.   W.M.P. van der Aalst. "Petri-Net-Based Workfow Management Software." *In Proc. NSF Workshop on Workflow and Process Automation in Information Systems*, Athens, Georgia - USA, May 6-8 1996. p. 114-118.

95.   W3C. "Uniform Resource Locators." 1994.
      http://www.w3.org/Addressing/URL/Overview.html

96.   W3C. "Overview of SGML Resources." 2000.

http://www.w3.org/MarkUp/SGML/

97. W3C. "Extensible Markup Language (XML) Activity Statement." 2003.
http://www.w3.org/XML/Activity

98. W3C. "HTTP - Hypertext Transfer Protocol." 2003.
http://www.w3.org/Protocols/

99. W3C. "HyperText Markup Language (HTML)." 2003.
http://www.w3.org/MarkUp/

100. W3C. "XML Schema." 2003.
http://www.w3.org/XML/Schema

101. W3C. "Web Services Glossary." August 2003.
http://www.w3.org/TR/2003/WD-ws-gloss-20030808/

102. W3C. "Extensible Markup Language (XML) 1.0 (Second Edition)." October 2000.
http://www.w3.org/TR/REC-xml

103. W3C. "Extensible Stylesheet Language (XSL)." October 2001.
http://www.w3.org/TR/xsl/

104. A.I. Wang. "Experience Paper: Using XML to Implement a Workflow Tool." *In Proc. 3rd Annual IASTED Conference on Software Engineering and Applications*, Scottsdale, Arizona - USA, October 6-8 1999.

105. A.I. Wang. "Using a Mobile, Agent-Based Environment to Support Cooperative Software Processes." PhD, Norwegian University of Science and Technology, Trondheim - Norway, 2001.

106. D. Wells. "Extreme Programming: A Gentle Introduction." 2003.
http://www.extremeprogramming.org

107. P.R. White. "Process Modelling and Computer Supported Cooperative Work." IOPT/35, Manchester University, Manchester - UK, April 18 1994. 21 p.
ftp://ftp.cs.man.ac.uk/pub/IPG/pw93a.zip

108. R.I. Winner, J.P. Pennell, H.E. Bertrand, and M.M.G. Slusarczuk. "The Role of Concurrent Engineering in Weapon Systems Acquisition." Report R-338, Institute for Defense Analyses, Alexandria, Virginia - USA, December 1988.

109. Workflow Management Coalition. "Terminology & Glossary." Document Number WFMC-TC-1011, Issue 3.0, February 1999. 65 p.
http://www.wfmc.org/standards/docs/TC-1011_term_glossary_v3.pdf

110. N. Zhu. "Pounamu - Model Things with Your Own Tool." Internal Report, Department of Computer Science, University of Auckland, Auckland - New Zealand, 2003.

Wissenschaftlicher Buchverlag bietet

kostenfreie

Publikation

von

wissenschaftlichen Arbeiten

Diplomarbeiten, Magisterarbeiten, Master und Bachelor Theses
sowie Dissertationen, Habilitationen und wissenschaftliche Monographien

Sie verfügen über eine wissenschaftliche Abschlußarbeit zu aktuellen oder zeitlosen
Fragestellungen, die hohen inhaltlichen und formalen Ansprüchen genügt,
und haben **Interesse an einer honorarvergüteten Publikation**?

Dann senden Sie bitte erste Informationen über Ihre Arbeit per Email
an info@vdm-verlag.de. Unser Außenlektorat meldet sich umgehend bei Ihnen.

VDM Verlag Dr. Müller Aktiengesellschaft & Co. KG
Dudweiler Landstraße 125a
D - 66123 Saarbrücken

www.vdm-verlag.de